HAS HAWKING ERRED?

HAS HAWKING ERRED?

Gerhard Kraus

With an introduction by
Professor Jan Boeyens

JANUS PUBLISHING COMPANY
London, England

First published in Great Britain 1993
by Janus Publishing Company
Duke House, 37 Duke Street
London W1M 5DF

Copyright © Gerhard Kraus 1993

British Library Cataloguing-in-Publication Data.
A catalogue record for this book is available from the British Library.

ISBN 1 85756 088 4

All rights reserved. No part of this publication may be reproduced, stored in a retrieval system or transmitted in any form or by any means, electronic, mechanical, photocopying, recording or otherwise, without the prior permission of the publisher.

The right of Gerhard Kraus to be identified as the author of this work has been asserted by him in accordance with the Copyright, Designs and Patents Act 1988.

Printed and bound in England by
Antony Rowe Ltd, Chippenham, Wiltshire.

Contents

	Note to Readers	vii
	Introduction by Professor Jan Boeyens	ix
	Preface	xiii
1	Hawking's Book – The Catalyst	1
2	The Role of Commonsense in Physics	4
3	The Principle of Regularity and Mathematical Vulnerability	8
4	Aspects of Relativity	16
5	The Concept of Time	19
6	Absolute Time	28
7	Can Time Slow Down?	34
8	The Concept of Space	49
9	Empty Space and Gravity	63
10	Space-time Analysed	66
11	Time as an Extra Space-dimension	77
12	The Laws of Transformation	83
13	The Lorentz Transformation	99
14	Space-time in Voluminal Space	105
15	The Enigma of the Curvature of Space	116
16	Velocity of Light and the Solar Calendar	131
17	Other Observations	134
18	Commenting on the Big Bang	141
19	Concluding Remarks	154

Figures

1	Portion of light visible to the human eye	6
2	The two-dimensional animal	52
3	The New York-Chicago Express	72
4	Space-weight diagram	79
5	Space-sound diagram	80
6	Space-temperature diagram	81
7	Lightbeam passing the sun	124
8	Paths of light from stars	136
9	Passage of light from stars	137
10	Deflection of light	138

Note to Readers

A. All my references to Stephen W. Hawking's *A Brief History of Time* refer to the Bantam Press 1988 edition. My references to Hawking are marked by the letter H along with the page number; both are bracketed. For example (H32) means: Hawking, page 32. What is important to stress here is that most of my references to Hawking's text are not verbatim quotations but are paraphrased in my own words. Verbatim quotations from Hawking's book are relatively few.

B. To give an authentic view of Einstein's ideas I have chosen Lincoln Barnett's *The Universe of Dr Einstein* (London, Gollancz, 1942) as my main reference source. References are marked by the letter B and the page number, both bracketed.

C. Other important references to Einstein are taken from his work *Relativity: The Special and General Theory* (first edition 1916). In 1952 Einstein supplemented the text with several additions. This enlarged edition was first published by Mather, London, in 1954. It was re-issued in 1960 as a University Paperback. My references are taken from the tenth edition of 1988. They are marked with the letter E and the relevant page number, both bracketed.

D. I draw on the wisdom of Roger Penrose, a close associate of Hawking, whose *The Emperor's New Mind* was published by Oxford University Press in 1989, and contains many challenging and original ideas on physics. References herein are taken from the Vintage

Has Hawking Erred?

Press edition of 1990. They are marked by the letter P, plus page number, both bracketed.

Introduction

by Professor Jan Boeyens,
Dean of the Faculty of Science,
Witwatersrand University,
Johannesburg

Science is a product of the human mind and only a minor one at that, considering the small percentage of thought devoted to it over the ages. It is therefore wrong to think that in some areas it has reached such abstract levels that it is no longer freely accessible to the layman. This only happens where confusion reigns and where it becomes beneficial to hide a lack of understanding under a cloud of obfuscation or tedious mathematics. Useful ideas in science are all simple and reducible to pedestrian discourse. By its very nature science has no final answers, no authorities and no enduring theories. It thrives by constantly challenging all postulates, conjectures, theories and hypotheses. It has no sacred laws, not even the second law of thermodynamics. Like the evolving human mind, science can never be static.

Science does have fashionable pursuits. Currently it is time, and this is no accident. Of all the parameters that feature in contemporary theories it is the most elusive and most difficult to analyse in a fixed frame. For this reason it has tended to be factored out of all formulae, producing the current body of time-free theories, no longer useful in the support of new thinking.

An unfortunate new complicating factor emerging into contemporary theoretical physics is the arrogant notion that final answers can at last be provided through consistent application of so-called standard models. There is a standard model for cosmology and a standard model for elementary particle theory – each one supporting the other. The symbiosis, however, is so cosy that no dissenting views can be tolerated. The fact that standard models provide no reasonable input into crucial debates around concepts like dark matter and anti-matter asymmetry, is simply ignored. The dogmatic appeal of the standard big-bang model has become so powerful that experimental observation is becoming a nuisance. Conflicting new data are simply neutralised by adding another untestable parameter, like an inflationary phase or a cosmic string, to the expanding universe. The interpretation of time as a physical concept becomes totally irrelevant. It takes the shape required by the model and could be either real or imaginary and go backwards as readily as forwards, without preference or logic.

Expecting the rampant theoreticians to pause at this point and reconsider the philosophical basis of important notions such as time, before rushing into complicated computations, seems to be in vain – a sad thought. The all-important arrow of time is not an ingredient of any standard model, but we hear that the end is in sight for theoretical physics. It probably means no more than the demise of standard models, without providing any further answers.

New understanding calls for some humility. The Copernican revolution did not start from a grand mathematical analysis, but rather from an effort to understand some obvious features of the solar system in a qualitative sense. Since that day it had to wait a long time for Kepler and Newton before it blossomed into the logical system that we know. The scientific understanding of the universe and of time has probably only now reached a stage of development comparable to the Copernican era for the solar system; a stage where it is necessary seriously to consider all novel ideas and new models. Who knows

Introduction

where the most useful new model will develop? It is very likely to be proposed by a rank outsider.

This is the spirit in which the present volume is to be read. There is a need for fresh ideas and this work consists of precisely that. It may be the first, but certainly will not be the last, challenge to the interpretation of Time in terms of current cosmology, which abounds with absurdities such as the big bang, many worlds, anthropic principles and universal title.

Preface

This book falls into the category of popular science. It not only aims to arouse the interest of the scientific specialist but more so that of the great mass of people of average intelligence who find fascination in such books as Stephen W. Hawking's bestseller *A Brief History of Time*. It requires no previous knowledge of the seemingly mysterious concepts of space-time or of Einsteinian relativity. Indeed, it is designed to burst the bubble of both these mysterious subjects.

The concept of space-time is now deeply entrenched in theoretical physics. The purpose of this book is to question the validity of this concept. Since the concept of space-time is the crux of Einstein's special theory of relativity (reaching into general relativity as well), it follows that the entire idea of Einsteinian relativity is thereby cast in doubt.

The arguments supporting this challenge were originally derived from a scrutiny of certain aspects of Hawking's book, which has figured for over three years on the world's non-fiction bestseller lists, having sold millions of copies. Hawking, confined to a wheelchair by motor neuron disease, has been called the greatest theoretical physicist since Einstein and holds Sir Isaac Newton's former chair as Lucasian Professor of Mathematics at Cambridge.

Has Hawking Erred?

What may surprise interested laymen and experts alike is that two such eminent scientists should be revealed as having adopted a principle (that of space-time) which, if the present book's claims are borne out, must now be considered obsolete; thereby **stamping space-time as one of the most significant misconceptions in the history of modern science.**

What needs emphasising here is that the idea of space-time presents only one specific aspect of Einstein's and Hawking's work and that rejecting it does not necessitate a rejection of the other far-reaching exploits that these two scientists have to their credit.

Whatever the case, I venture to predict that the opinions advanced in this book will be hotly contested by sections of the physics establishment, notwithstanding the fact that the arguments advanced in support are based on solid evidence. Considering on the other hand the complexity of the subject under review, it is likely that some weaknesses in the detailed presentation may have slipped in, leaving the argumentation open to criticism. I predict, however, that such nitpicking *per se* will not be able to damage the core of my argument, but leave its challenge intact.

Some objections have been raised that throughout the book there is much repetitive argument. My response is, that I find such repetitions necessary in order to illuminate the principles of space and time in the different contexts in which they appear in the different chapters.

1
Hawking's Book – The Catalyst

For a long time I had harboured doubts about certain mind-boggling views held by Einstein and his modern successors: that time can slow down, come to a standstill and even turn backwards; that three dimensional voluminal space can accommodate additional space dimensions, one of them being time, resulting in space-time; and that voluminal space and time can warp and bend. I began to make notes on these points, which to my mind seemed to violate the most basic assumptions of commonsense.

It was the reading of Hawking's bestseller which then persuaded me to set out my observations about these problems, resulting in the present book. I did this despite the fact that it meant contradicting such formidable authorities as Einstein, and of course Hawking, on their own territory, when I myself was not a specialist in the physical sciences. Perhaps my endeavour might thus be construed as that of a fool rushing in where angels fear to tread.

I must emphasise that my evaluation of Hawking's book does not mean that I am questioning the entire range of his prodigious work, which is far beyond the scope of the present treatise. Thus my views present only a restricted criticism, referring specifically to the subjects mentioned above.

Has Hawking Erred?

It has been claimed that Hawking's bestseller, although tailored for a readership not necessarily versed in the intricacies of modern physics, presents a brilliant summary of the laws governing the universe. The astronomer Carl Sagan, in his introduction to Hawking's book, said among others that this work is as interesting for the insight it conveys into Hawking's thinking, as it is for its far-reaching contents. There are new insights on the front edges of both science and personal courage.

While appreciating Hawking's book (and his other admirable contributions to modern physics), I believe that I have found several specific contradictions in his text which throw doubt on the validity of the four dimensional space-time concept. This concept was adopted by Einstein several decades ago and has since become something of a sacred cow to most physicists the world over.

Despite such universal acclaim, I have become convinced that by uncritically accepting Einstein's space-time theories, which play such a large part in theoretical physics, Hawking, in common with other physicists, may be perpetuating a major fallacy unique in scientific history. Moreover, though he calls his book *A Brief History of Time*, I find that Hawking has failed to present a clear definition of the concept of time.

In contrast to the brilliant genius of Hawking, I admit that my command of physics and mathematics is of a lower order, stemming from my days as a student in a technical college, supplemented by additional research over the last several years. What I like to emphasise in this respect is that my approach to Hawking and Einstein is motivated by a healthy scepticism. I am merely taking a leaf from Einstein himself, who, according to Barnet (B49), proclaimed that he was unwilling ever to accept any scientific principle as self-evident, merely at its face value. To me the principle of space-time, as it is postulated by Hawking and Einstein, is far from self-evident.

Hawking's Book – The Catalyst

Before starting on what appears a rather daring enterprise, I will work up a background with an elucidation on three specific subjects, which appear to me an essential prelude for the understanding of the argument to follow. They are: 1. The role of commonsense in physics; 2. The principle of regularity and mathematical vulnerability; and 3. Aspects of relativity.

2
The Role of Commonsense in Physics

In his book *Superforce* (1984), physicist Paul Davies includes a chapter entitled 'The New Physics and the Collapse of Commonsense'. He maintains that 'the New Physics' defy almost all notions of commonsense, transporting the layman into an Alice in Wonderland world. Davies points out that reality is nevertheless preserved by the wondrous feats of mathematics, which to his mind rule all physical events. At a later stage he goes as far as finding in this physical labyrinth, bolstered by mathematical equations, some striking similarities with the mysticism of Indian and Chinese philosophy. He writes (p. 220):

> Writers such as Fritjof Capra in the *Tao of Physics* and Gary Zukav in *The Dancing Wu Li Masters* have emphasised the unity of existence and the subtle relationship between the whole and its part

To me, such a total rejection of commonsense pushes modern physics into the realms of metaphysics.
Einstein points to this trend when he says (B12–13):

> In accepting a mathematical description of nature, physicists have been forced to abandon the ordinary

The Role of Commonsense in Physics

world of pure experience, the world of sense perceptions. To understand the significance of this retreat it is necessary to step across the thin line that divides physics from metaphysics. Questions involving the relationship between observer and reality, subject and object, have haunted philosophical thinkers since the dawn of reason.

Hawking (H18) maintains that while commonsense notions may well apply to such earthly objects as apples and even to our comparatively slow-moving planetary system, they fail to work when we have to deal with objects that move at or near the speed of light.

This is only superficially true, because it can be shown that the notion of 'commonsense' has a very wide-ranging meaning. At least three different aspects of it can be enumerated.

(a) Plain or ordinary commonsense

This is based on normal sensory experience and its limitations, aptly illuminated in an example cited by Barnett (B14–15), that deals with our faculty of vision. He explains:

> Anyone who has ever thrust a glass prism into a sunbeam and seen the rainbow colours of the solar spectrum refracted on the screen has looked upon the whole range of visible light. For the human eye is sensitive only to a narrow band of radiation that falls between the red and the violet. A difference of a few one hundred-thousandth of centimetres in wavelength makes the difference between visibility and invisibility . . . It is evident, therefore, that the human eye suppresses most of the 'lights' in the world, and what man can perceive of the reality around him, is distorted and enfeebled by the limitations of his organ of vision. The world would

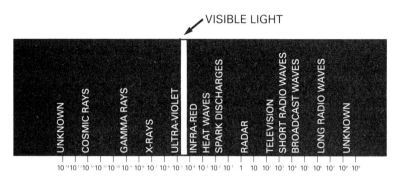

Fig. 1: Portion of light visible to the human eye

appear far different to him if his eye were sensitive, for example, to X-rays.

The fact that physicists have been able to overcome the frailty of our inborn biological range of vision is entirely a result of using our commonsense in an ingeniously orchestrated way, leading us to what I call an expansion of commonsense. Thus:

(b) Expanded commonsense
the range of normal sense perception, enlarged by means of instruments and mathematical calculations.

(c) Abandonment of commonsense
This involves the virtual rejection of ordinary sensory experience and the transformation of physical events into metaphysical phantoms.

I venture to maintain that the science of physics cannot abandon the concept of 'ordinary commonsense', for the simple reason that even the most intricate scientific instrument and the most complex mathematical equation is ulti-

The Role of Commonsense in Physics

mately dependent on normal sensory experience, or what is more obliquely called 'the human mind or intellect'.

The medieval paragon of science, Paracelsus, once said that while God may be able to create an ass with three tails, he cannot create a triangle with more than three sides. Hawking and Einstein, by creating voluminal space of more than three dimensions (including space-time), and going further into five, six, ten and more space dimensions, have created the equivalent of more than three-dimensional triangles; thus, in the Paracelsian sense, they have outsmarted God.

'Plain ordinary commonsense' is the ultimate source of all wisdom and forms the fountainhead from which all scientific enquiry emanates; with the proviso that it must be handled with careful consideration and based on ascertainable facts.

3
The Principle of Regularity and Mathematical Vulnerability

One factor in science, the importance of which seems to have hitherto not been given the credit it deserves, is the 'principle of regularity'. Its role, not only in physics but in science as a whole, easily exceeds that of relativity. Firstly, all measurements of time, space, temperature, weight, sound and gravity, are geared to regular units and processes. Secondly, all mathematical formulae, whether relating to abstract quantities or real physical ones, are based on the regularity of numbers. Any law in physical science only gains validity if dealing with quantities, interacting on the basis of regular units of measurement. Thus when Einstein wrote (B38) that 'the laws of nature are the same for all uniformly moving systems', he hit the nail squarely on the head. Regretfully, though, he deviated in his later work from this cast-iron rule when he dealt specifically with the measurement of space and time, adopting the questionable axiom (B46) that 'time and distance are variable quantities'.

Regularity, of course, implies order. Believing in an ordered universe (which is obviously one where the laws of nature interact in a regular manner), Einstein expressed his belief in the famous statement 'God does not play dice'. However, this saying, to strip it of any metaphysical

The Principle of Regularity and Mathematical Vulnerability

meaning, must be taken in a metaphorical sense. Even the irregularities discovered by Heisenberg in sub-atomic particles become somehow transformed into regularities by incorporating them into the regular equations of quantum mechanics. As to the uncertainty principle of Heisenberg itself, both Einstein and Hawking expressed the opinion that there may be physical laws, as yet undiscovered, which might transform present sub-atomic uncertainties into measurable certainties.

Hawking's comment on this particular aspect of particle physics elucidates that the 'uncertainty principle' as inherent in quantum mechanics deals with events occurring on a sub-atomic level. It postulates that conventional measuring methods fail to pinpoint the position of particles and their velocities simultaneously and with certainty.

However, this apparent unpredictability may only occur when attempts are made to interpret waves in terms of positions and velocities of particles. Hawking suggests that this approach can be mistaken. Perhaps there are neither positions of particles nor velocities, but only waves? And perhaps we try to fit the waves into preconceived ideas about velocities and particles? The resulting misinterpretation, concludes Hawking, may be the cause of an apparent unpredictability.

Using the principle of regularity, events in nature can be synchronised with the decimal system of mathematics. Thus, through having invented a system of numbers which reflects such regularity, humankind has designed a means to unlock some of the most concealed forces of nature. One result was Einstein's famous equation $E = mc^2$, a formula which discloses the awesome powers hidden in the tiny atom.

Similar considerations have persuaded present-day physicists to treat mathematics as an almost infallible medium for revealing the secrets of the physical world.

Has Hawking Erred?

One physicist, Paul Davies (*Superforce*, 1984), invests mathematics with an almost religious awe. The roots of such thinking go back into antiquity. Greek mathematical and astronomical ideas are shown to have derived from Babylonia and Egypt. It is also believed that the Pythagorean theorem was not an original invention of Pythagoras but was rediscovered by him through his contacts with Babylon, where it was known as early as 2000 BC. On the other hand, Herodotus in *The Histories*, reports that Pythagoras learned numbers and measures from the Egyptians. The archaeologist Gordon Childe writes (*Man Makes Himself*, 1966):

> We do not know how any of the geometrical rules were obtained. They certainly had not been deduced *a priori* to be in Euclid's geometry. For such a 'pure geometry' there is no evidence at all. Figures often accompany the geometrical problems in the mathematical papyri and tablets, but such figures are no more drawn to scale than are the fieldplans in contracts. On the other hand, the patterns decorating craft products, piles of bricks and composite wooden boxes often give very striking colour demonstrations of geometrical propositions . . . Fourteen hundred years later the physical origin of such geometrical abstractions as well as those of numbers and mathematical formulae had become obscure and forgotten. The mystery relationships between squares and circles, between squares and triangles, etc. seem to have been taken over by Pythagoras, as well as Aristotle, Plato and others, as philosophical truths.

Roger Penrose, referring to such traditions (P554), writes: 'According to the Platonic viewpoint mathematical ideas have an existence of their own'.

In other words, all mathematical concepts exist already, concealed within nature's events. The mathematician's

The Principle of Regularity and Mathematical Vulnerability

task is to discover them, that is, express physical events in mathematical formulae. Penrose observes (P225):

> It is remarkable that **all** the SUPERB theories of nature have proved to be extraordinarily fertile as sources of mathematical ideas. There is a deep and beautiful mystery in this fact: that these superb accurate theories are also extraordinarily fruitful simply as **mathematics**.

And (P554) that 'according to the Platonic viewpoint mathematical ideas have an existence of their own'. Referring to the system of complex numbers, Penrose further remarks: 'With such identification, it might be more comprehensive how "minds" could seem to manifest some mysterious connection between the physical world and Plato's world of mathematics'.

However, human beings are far from perfect, and we do not live in a perfect world. Hence mathematics, although producing phenomenal results, but being a human invention, cannot avoid the pitfalls of human frailty. It is sometimes maintained that mathematical capacity is an inborn property, being one of the latest biologically evolutionary acquirement of modern Homo sapiens, seated in a specific part of the human brain. Evolutionary biology tells otherwise.

Latest studies of human evolutionary history have revealed that Homo sapiens appeared on this earth sometime between 400,000 and 300,000 years ago, following the species Homo erectus in the evolutionary scale. What is significant is that ancient Homo sapiens can be shown to have had the same average brain volume (i.e. 1350 cc) and the same intellectual qualities and potential as contemporary modern man, who has more or less biologically stagnated over the entire period. This means that humans of 300,000 years ago could have theoretically pro-

Has Hawking Erred?

duced the genius of an Einstein or that of any other prominent mathematician, had the opportunity offered itself.

It was not until between 4200 and 4100 BC or so, that the invention of writing associated with the use of numbers came into existence in the wake of the invention of a calendar, releasing thereby hitherto dormant mathematical capacities – though initially for purely practical and ulilitarian purposes. Gordon Childe writes (*Man Makes Himself*, 1966):

> It must not be supposed that ancient societies were interested in infinite length or empty space. Their abstractions were limited by practical interest. The ancient Sumerian measures of area have in some cases the same names as measures of weight; in particular the smallest unit in both 'tables' is the Se or grain. In other words, Sumerian 'square measure' was originally a seed measure. The Sumerian interest was the quantity of seed needed to sow the fields. He regarded the field not as occupying so much empty space but as needing so much seed grain. With the areas of bits of uncultivatable desert or blue sky he was not concerned.

Theoretical, abstract mathematics with all its modern expressions is a descendant of these rudimentary beginnings.

Despite its phenomenal achievements, modern mathematics cannot claim perfection, being based on the frailties of the decimal system. For example, man has never succeeded in squaring the circle, being unable to produce a conclusive number for pi. Thus in reading 3.14159265358979323846 ad infinitum must forever remain inconclusive in mathematical terms, although the relationship it intends to express, namely that of a circle's radius to its surface and to its circumference, is definite and conclusive. The same applies to Planck's constant of quan-

The Principle of Regularity and Mathematical Vulnerability

tum mechanics. Further, it can be shown that theoretical mathematics, as projected on paper, does not always coincide with natural events.

Einstein, on occasion, speaks of the stern logic of mathematics (B43), as it is inherent in the simple arithmetical equation of two plus two equalling four ($2 + 2 = 4$). It can be shown, however, in the following, that this is only partially correct, and is not applicable to all events in the natural world. For instance, adding one droplet of water to another one will again only give one droplet (though a bigger one), resulting in the relative arithmetical equation of one plus one equalling one ($1 + 1 = 1$). This purely numerical abstraction pays no heed to either size, form or the nature of objects. Going further and uniting two droplets of water with another two, will result again in only one droplet (though this time a still bigger one); the result being $2 + 2 = 1$. In the case of fluids or other suitable materials like lumps of plasticine rolled into spheres, etc., the arithmetical formula of $1 + 1 = 1$, also equalling $2 + 2 = 1$, can result in the paradoxical equation of $2 + 2 = 1 + 1$ or 4 equalling 2, or in many other equally absurd combinations.

What the preceding example teaches us is that other purely abstract mathematical calculations made on paper may, when we try to relate them to practical physics problems, result in mathematical formulae which have no relation to physical reality. This is amply confirmed in the mathematics of space-time. The latter, as will be shown subsequently, having no basis in reality, can therefore not produce any valid mathematical calculations to make its expectations come true.

Furthermore, the possibility that certain Einsteinian equations may not necessarily match natural events, (which they are supposed to anticipate), was recently demonstrated by two researchers at Cornell University, Stuart Shapiro and Saul Teukolsky, professors of astronomy and physics. They used a supercomputer to simulate a tremendous gravitational collapse in the universe, producing states of matter, an event that according to Einstein's general theory of relativity was impossible (see *Physical Review Letter*, 25 February 1991).

It had not been established until then whether **Einstein's equations** could give rise to naked singularities not clothed in black. Dr. Shapiro said: **'We don't believe naked singularities actually happen in nature.** So if the theory says that they are occurring that's the breakdown of the theory' [emphasis supplied]. Other scientists who studied the results came to similar conclusions, namely, that the simulation may have exposed a flaw in at least one aspect of Einstein's theory on the behaviour of space, time, matter and gravity.

Another point of the same order concerns extra space dimensions. Here attempts abound to prove mathematically (though in doing so only on paper) the existence of a great number of extra space dimensions. Paul Davies (in *Superforce*, 1984, p. 237), cites an example where an eleven dimensional space feature emerged from lengthy mathematical analysis. Hawking (P162), writing about 'string theories', points out that they seem to be consistent only if space has either ten or twenty-six dimensions instead of the usual four.

As far as Einstein is concerned, we can indicate here in advance of later proof that the same fate of mathematical fallibility enshrouds the Lorentz equations, which play such an important role in his endeavour to establish that time and distance are variable quantities. Einstein asserts (B46) that:

> The scientist who wishes to describe the phenomena of nature in terms that are consistent for all systems throughout the universe must regard measurements of time and distance as variable quantities.

The Principle of Regularity and Mathematical Vulnerability

He asserts that:

> The equations comprising the Lorentz transformations do just that. They preserve the velocity of light as a universal constant, but modify all measurements of time and distance according to the velocity of each system of reference.

This, in other words, amounts to a rejection of 'regular calendar time' as the all-embracing medium, which, with the help of clocks or other suitable timepieces, commonly measures the duration of all events in the universe, including the velocity of objects. The problems involved in this proposition are extensively dealt with in chapter 5 The Concept of Time.

Finally, to conclude my remarks on mathematical vulnerability, I quote from an article by US physicist John Lukacs, printed in the *International Herald Tribune*, 18 June 1993. The author, commenting on the proposed giant Texan Supercollider, writes:

> More and more mathematical formulae about subatomic matter consist only of untested and untestable assumptions, all of them theoretical and abstract. The belief that the universe is 'written in the language of mathematics' is not only wrong, it is entirely outdated. 'What is there exact in mathematics except its own exactitude?' Goethe wrote. He was right, as mathematicians in the 20th century have confirmed.

4
Aspects of Relativity

All manifestations in our natural world or, in a wider perspective, in the entire universe, find practical expression only when related to tangible entities. This applies both to organic life and to the inorganic mechanical processes of measurement. In organic life all sensory perceptions relate to sense organs – sight to the eyes, odours to the nose, taste to the mouth, hearing to the ears and skin sensations to the perception of touch, and so on. In the inorganic world all measurements, whether of temperature, weight, sound, density, space (linear and voluminal) and of time, are based on, and related to, arbitrarily fixed frames of reference.

For objects in motion it is necessary to point out that there exists no fixed point in the universe to which measurements of time and position can be related in an absolute fashion. Hence, ever since scientific calculations began, all measurements referring to time, space, temperature, weight, etc., were related to some arbitrarily fixed frame of reference. For this very reason relations between two freely moving systems, although they can be set down in terms of mathematical formulae, are of little practical value and must remain fleeting assessments unless they are related to some extraneous, arbitrarily fixed frame of reference of a third order.

Aspects of Relativity

For the latter reason, physicists have generally used our earth or, in a wider aspect, our solar system, as pivots for physical measurements. Thus, all valid human time measurements in our universe are, without exception, based on our conventional terrestrial calendar and its subdivisions; while all linear space measurements are based on our common metric system. Both are founded on criteria which will be outlined later on. In this, our practical application of the measurements of time, distance and volume, no allowance can be made for the slowing down of clocks or the shrinkage of measuring rods, which in the theory of relativity are postulated to occur when clocks or measuring rods are presumed to accompany objects moving at great speeds. As a rule, real, practical measurements are usually arrived at indirectly by calculations on paper. Examples are Hubble's contention that a light ray circling the universe would take 200 billion terrestrial years to accomplish this feat; and that the sunlight takes eight minutes to reach our earth. And although Einstein and his disciples have theoretically proved that measuring rods will shrink and clocks will slow down when accompanying objects travelling at great speeds, this exercise is merely of theoretical value and cannot be seen to have any practical application.

It is noteworthy that most of the above observations on relativity and their interpretation accord with Einstein's own ideas on the subject. For instance, Einstein points out (B40):

All the clocks ever used have been geared to our solar system. What we call an hour is actually a measurement in space – an arc of 15 degrees in the apparent daily rotation of the celestial sphere. And what we call a year is simply a measure of the earth's progression in its orbit around the sun . . . Relativity tells us there is no such thing as a fixed interval of time independent of the system to which it is referred. There is indeed no such thing as 'simultaneity', there is no such thing as 'now', independent of a system of reference [emphasis supplied].

Further on (B41) he states:

It is constantly necessary for the scientist, in dealing with matters involving complex forms of motion (as in celestial mechanics, electrodynamics, etc.), to relate the magnitudes found in one system with those occurring in another.

To this I must add the proviso that one of the two systems (or an arbitrarily designed third system) must serve as a fixed frame of reference. This is again confirmed by Einstein (in *Relativity*, 1988, p. 7): 'If the results of physics and astronomy are to maintain their clearness, the physical meaning of specification of position must always be thought with the above considerations' [**i.e. in reference to a fixed system of coordinates, G.K.**]. Hence Einstein: 'Every description of events in space involves the use of a rigid body to which such events have to be referred'. To this he adds (B 33): '**For all ordinary purposes of science, the earth can be regarded as a stationary system**' [emphasis supplied].

While all this conforms with our own views on relativity as set out above, it is remarkable to note (as it will be subsequently shown) that when Einstein deals with the phenomenon of light, he appears to exempt it from the general principles of relativity as shown above, by implying that its proclivities involve a 'deep enigma of nature' (see B43). Einstein elaborates this exemption when dealing with the mathematical laws underlying the relationships known as 'laws of transformation'; these are considered in a later chapter.

5
The Concept of Time

I believe that the phenomenon we call 'time', specifically when used in science, only gains an unequivocal, clearcut meaning when used as a **comparative measure**. Apart from this, it appears that the purely abstract use of the time concept defies a clear definition and somehow strays into the realms of abstract philosophy, like the proposition that physical existence presupposes time by differentiating events in their relative succession. One offshoot of the time concept is 'absolute time', which seen from a cosmic viewpoint lacks the prerogative of reality, but which in Hawking's and Einstein's definition (see chapter 6) inadvertently identifies with conventional calendar time. The seemingly commonsense notion that there is something like an inborn time sense in humans is also questionable.

Penrose (P391) asserts that 'central to our feeling of awareness is the sensation of the progression of time'. This would presuppose the existence of an additional sense-organ with a 'time sense'. And although the succession of events in the universe and their relative progression signifies life, being and existence, as distinct from non-existence (with existence being seemingly time-linked in a philosophical sense), this presumed capacity is not biologically imprinted on the organism. Seen from a biological viewpoint there is no such thing as a sense organ in either humans or animals which consciously perceives time. The experience of cave explorers who were lost for days in dark caves, cut off from light and sound, and who consequently completely lost any track of the progress of time, has been reported. Equally so, blind people in isolation from sounds and any other sensory experience become completely disorientated as to time perception. Thus, the experience of time seems to be entirely linked to the passing of perceptible events. In practice, it gains reality only when their duration is measured relative to conventional calendar time.

The nearest phenomenon which might be described as a 'time sense' (which it is not), is an unconscious reflex action of the body called the 'circadian clock' or 'rhythm'. (Circadian – from Latin 'circa' = approximate, and 'dies' = day.) It is a genetic adaptation of environmental influences such as temperature, climate, the day/night cycle, etc., causing changes in body temperature, blood pressure, the heart beat, hormonal changes and so forth. This biological property is prevalent in humans, animals and plants and cannot be classified as a 'time sense'.

Time measuring:

Thus, in its everyday practical use 'time', like the metric measure, is not more and not less than a comparative means, which has no intrinsic existence of its own (as it has in the purely abstract expression of 'absolute time'). It is a specific human invention, tailored to compare the intervals between passing events, the speed of moving bodies, the duration of such movements, the ageing of objects both in motion and at rest, as well as the ageing of growing organisms, and overall the age of the universe in which all these events occur, have occurred in the past and, it is presumed, will occur in the future.

Furthermore, all these happenings are attuned and measured relative to our conventional calendar year, itself

The Concept of Time

divisible into months, weeks, days, hours, minutes, seconds and their subdivisions. To carry out time measurements, timepieces have been devised which, to be of any practical use, must work in strict synchronisation with our earth/sun relationship, both for ordinary everyday use and for scientific purposes.

It is important to note here that Einstein based time measuring originally on the same premise, although he later deviated from this most axiomatic fact: 'ALL THE CLOCKS EVER USED BY MAN HAVE BEEN GEARED TO OUR SOLAR SYSTEM' (B40).

From a historical viewpoint, time measurement began with the invention of a calendar in pre-dynastic Egypt. It is presumed that it was based on observing the fairly regular arrival of the Nile flood in Lower Egypt. Its prediction became a factor of vital importance for the agricultural pursuits of the Nile Valley dwellers – and thus became a basic element in the origin of Egyptian civilization. In the Cairo area (before the Aswan Dam was built), the Nile normally rose 23 feet. By the end of September the entire Nile area was submerged and presented the appearance of a large sea of turbid waters. Settlements built on high ground rose like islands. For a month this water remained stationary, before subsiding rapidly.

The arrival of the Nile flood was always awaited with great anxiety. The archaeologist Gordon Childe, has written (*Man Makes Himself*, 1966):

> Now it happens that just about the time the flood should reach Cairo, the last star to appear on the horizon before dawn obscures all stars, was Sirius, the Egyptians' Sothis. The helical rising of Sirius was thus a naturally fixed point in the solar year. It was found that this event recurred roughly every 365 days, and it was then taken as the starting point of an artificial State year of that length. The Egyptian calendar was accord-

ingly divided into 12 months of 30 days each, plus five days extra and an additional day every four years.

Whether a lunar calendar actually preceded the Sothic calendar has never been resolved. On the other hand, it has been suggested (though there is some doubt about the dates) that a more accurate Egyptian calendar was based on the Sothic cycle and that its beginnings can be put at between 4241 and 4240 BC. These dates mark the presumed (though not confirmed) inception of the Egyptian calendar. The strictly astronomical aspect of these dates has been briefly summarized by Bray and Trump in the *Penguin Dictionary of Archaeology*, 1972. They reiterate that the civil and calendar year was based on a year of 365 days rather than the correct 365.422 days, hence lagging steadily behind the true solar year. The length of this cycle was 1,460 years to the time when the two calendars were again exactly in step; this did happen, according to a record of Theon of Alexandria, in AD 139.

As to the modern use of the conventional 'Solar Calendar' the following details are relevant. Two basic kinds of time are now maintained: International Atomic Time and Universal Coordinated Time. International Atomic Time is derived by the International Bureau of Time at the Paris Observatory from three cesium clocks of extreme precision in Canada, West Germany and the United States. Such time runs continuously and is used, for example, to match astronomical observations from different continents or to start up widely separated power-generating systems in phase with one another. The cesium clock at the US Naval Observatory in Washington, like all cesium clocks, churns away the hours at the rate of 9,192,631,770 oscillations per second. Time signals from national services are monitored in Paris, and once a month each service is told if its clocks are running fast or slow.

Universal Coordinated Time, while it runs at the same

The Concept of Time

rate, is designed to keep in phase with the earth's rotation. This is the standard that the everyday world lives by and that governs time announcements on radio or over the telephone. It differs from time as it might be determined by astronomical observations in that such predictable effects as wobbles of seasonal variations in the earth's spin are averaged out. To demonstrate the exclusive use of regular conventional calendar time in all spheres of astronomy and physics, I am citing a number of typical examples, in which I emphasise all references to the regular units of calendar time, such as seconds, minutes, days and years.

Einstein mentions (B41) that the star Arcturus is 38 light **years** away from us; also that the nearest galaxies are estimated at 100 million light **years** away from us and the remotest about 500 million light **years** away. The astronomer Hubble, using Einstein's field equations, has calculated that the radius of the universe measures 35 billion light **years** and that a sunbeam setting out through space at the rate of 186,000 miles per **second**, would encircle the universe and return to its source after a little more than '**200 billion terrestrial years**' (B85/86). Unless the above time measurements are based on a perfectly regular time-flow related to the earth-sun cycle, they would make no sense.

Hawking (H24), notes that a sunray takes four **years** to reach the star Alpha Centauri; also (H37), that we live in a galaxy which orbits around its centre about once every several hundred million **years**, and further (H46), that between 10 and 20 thousand million **years** ago the distance between neighbouring galaxies must have been zero. He consequently concludes (H108) that the universe is between 10 and 20 billion **years** old. By thus using the principle of 'regular calendar time' and its subdivisions in the above context, Hawking implies that all events in the universe since its suggested beginnings, whether they

Has Hawking Erred?

relate to the large galaxies of stars or the minutest particles of matter, took place within a time framework of about 10 to 20 billions of **years**. These **years** are expressed in conventional solar, or calendar **years** of a regular time flow, each again counting about 365¼ **days** and further divisible into **conventional hours, minutes, seconds and picoseconds**.

The Big Bang theory of the origin of the universe holds that after the initial explosion of the superhot ball that began it all, a process started that led to the creation of atoms within 500,000 **years** and the early formation of stars and galaxies 200 million years later. After about 300,000 **years** from the event, the opaque plasma from the primary explosion turned into more transparent gas, resulting in microwave radiation (*International Herald Tribune*, 15 January 1990, reporting on the unmanned spacecraft 'Cosmic Background Explorer'). All this only makes sense on the assumption that **all the years** mentioned follow each other in perfectly regular sequence. Furthermore, the **years** in question are those of **regular terrestrial calendar years**, which for all practical reasons are based on units, which we can justifiably describe as **regular calendar years**.

Eric J. Lerner, who disputes the 'Big Bang Theory' (*International Herald Tribune*, 3 June 1991), points out that 'Because galaxies travel at small fractions of the speed of light, mathematics show that such large clumps of matter must have taken at least **100 billion years** to come together'; thus, if correct, outdistancing Hawking's assumption of a 10 to 20 billion **year** old universe five times over. My point here is again that **the 100 billion years in question, if applicable, are conventional terrestrial calendar years**.

A theory postulating that the cosmos is **trillions of years** old furthermore disputes the 'Big Bang'. Such a time scale only make sense when it is assumed that **each of these**

The Concept of Time

billions or trillions of years is of equal duration, proceeding throughout the entire period as the same uniform pace. This uniform progress of what is obviously 'regular calendar time' must prevail, whether the time scale is based on the conventional solar calendar or on any other possible alternative time relationship, attuned to, or synchronised with, conventional calendar time.

It involves the same principle which rules that a distance measured between two points remains unchanged, whether we express it in inches or centimetres, miles or kilometres. For example, Mercury orbits the sun in 88 of our days, and in the same period rotates just once on its axis. So the duration of a Mercury year and a Mercury day amounts to the same thing. A regular Mercury year equals 88 earth days and a regular Mercury hour equals 88 divided by 24, or 3.66 earth days.

Adhering to the same, regular calendar time scale, particle physics uses tiny time fractions like **microseconds** and **picoseconds**, the latter being a mere 10^{-12} part of **a second of our regular terrestrial time standard.** Furthermore, picoseconds, which are a trillionth of a second, are still divisible into femtoseconds. For example, there are more femtoseconds in one second than there were seconds in the past 31 million years.

Referring to the creation of the universe, Penrose, an adherent of the 'Big Bang Theory' points out (P424–5)

> From that moment, **one thousandth of a second** after creation (i.e. 10^{-4}), the emergence of the primordial fireball, until about **three minutes** later, can be described by well-established physical theory . . . Thus it was the initial fireball that spread its gas so uniformly throughout space [emphasis supplied].

Among other phenomena related to conventional calendar time is the **'light second'**. According to Hawking

(H23), **a light second** is simply defined as the distance **the light travels in one second (of terrestrial time)**, this being some 300,000 kilometres (186,284 miles). One light metre, again being the 300,000 millionth part of a **light second**, has been accurately measured by cesium clocks and equals in duration 0.000000003335640952 parts of **a conventional terrestrial second**. Its multiple is convertible into **conventional terrestrial minutes, hours, days and years**, which gain their more accurate expression in 'sidereal time'. A sidereal day is the period of a complete rotation of the earth upon its axis with respect to the fixed stars, while the sidereal second is the 24th part of a sidereal day, divided by 3600. Furthermore, 'regular calendar time', as related to sidereal time, can be synchronised with the speed of light, another absolute constant in the universe.

Other phenomena of nature which are **strictly related to our conventional terrestrial second** are Einstein's famous formula for energy and the 'Planck's Constant' of quantum theory.

In Einstein's formula $E = mc^2$, E denotes energy m denotes mass and c stands for the velocity of light which proceeds at a pace of 300,000 kilometres **per terrestrial second** anywhere in the universe.

The German physicist Max Planck discovered that electromagnetic oscillations occur in quanta, where energy E has a definite relation to their frequency v, given by the formula $E = hv$, with 'h' being recognized as being very tiny, equalling 6.6×10^{-34} 'Joule seconds'. A 'Joule second' again is a unit of work, equalling 10^7 ergs, and equals the work done in **one second** by a current of one ampere flowing through a resistance of one ohm. What needs stressing here is that a **'Joule second' is identical in duration to our conventional second**, thus forming a basic ingredient of quantum mechanics.

It is evident that **physicists the world over (including Einstein and Hawking) have measured and expressed all** macroscopic and microscopic events in the universe in terms of regular terrestrial calendar time, which accords with the light second (measured in terms of conventional time). Thus they use, as their universal time standard, the main aspect of what Einstein (as shown later) defines as 'absolute time'.

6
Absolute Time

The preceding examples show convincingly that all practical time measurements in astronomy and in physics (though not necessarily their theoretical accompaniments), as well as those of everyday life, are based on our conventional terrestrial calendar. It has further been shown that this is based on the universally applicable criteria of the sun/earth relationship. What needs stressing here is that the resulting time readings also apply to events which happened billions of years before our own planetary solar system came into being. More remarkable still is the fact that this time measurement also applies **retroactively** to the tiniest fractions of seconds, down to a primordial fireball's appearance one thousandth of a second after the creation of the universe. As to macrocosmic events I refer to Hawking's assumption that the universe is between 10 and 20 billion years old. What is notable here is that such an assumption can only hold true when it is also assumed that all the 10 to 20 billion years in question were of equal duration, with each year following another in perfectly regular sequence down to the smallest split second. Hawking's assumption, it would appear, must therefore exclude any possibility of embodying in its system any kind of time irregularity.

I reiterate therefore that time measurement based on

Absolute Time

our terrestrial calendar presents a system of absolute regularity throughout the history of the universe, disallowing any kind of time slowdown, or even a possible standstill of time as postulated in the time theories of Hawking and Einstein.

Furthermore, the linear distance measurements accompanying all events in the universe enumerated above are based on the unit of our **standard metre**, whose prototype is projected between two markings on a platinum bar lodged in Paris, denoting the metre's length. In the following I am taking astronomer Hubble's postulate that a light ray encircling the universe would take 200 billion years to do so, as an example for metric measurement. To measure the distance the light ray travels during this time we have to translate it into metric units as follows. Firstly, one light year, which is the distance light travels in one year, has to be converted into light seconds. Accordingly, light travelling 300,000 kilometres per second has to be multiplied by $365¼ \times 24 \times 3600$ to get the number of kilometres it travels in one year. The result is 9.4 trillion km (equalling 5.9 trillion miles). This figure is again to be multiplied by 1000 (1000 metres being equivalent to one km). By multiplying the resulting distance again by 200 billion, we obtain the total metric distance the light will have covered during its journey circumscribing the universe. I am giving this example to show that our common metric unit of measurement is uniformly applicable throughout the universe, without allowing any variation in its length, and that it can even be retroactively applied over billions of years, although its introduction occurred only recently.

These details indicate that our conventional time and distance measures are quite absolute insofar as they are universally applicable and are retroactively valid back to the creation of the universe. This, according to Hawking, occurred 10–20 billion years ago. One may therefore justifiably describe our time and metric measures as expressing absolute values, being reflected in such phrases as 'absolute time' and 'absolute distance'.

There are, however, two aspects of 'absolute time':

    a) The abstract notion of an all-pervading God-given time factor, denoting a purely abstract entity, and

    b) Our notion of absolute time, strictly based on our solar system.

Remarkably though, this latter definition of 'absolute time' matches both our own notion of 'absolute time' and that subsequently elucidated by Hawking and Einstein.

Defining the term 'absolute'

In first trying to define the meaning of the word 'absolute', as used in science, I admit that, theoretically viewed, there cannot exist any completely 'absolute' values in our ever-changing universe. The practical use of the term 'absolute' is therefore conditional on the system of reference to which it is related. For example, chemistry has a formula for 'absolute alcohol', which is related to ethyl alcohol of not less than 99% purity. Significantly though, a 100% purity is not conditional. 'Absolute zero', which denotes the total absence of heat energy, is fixed at 273.16° C below zero. Although this figure is calculated in relation to earthly measurement conditions, which are variable, it is valid throughout the entire universe. In our case, the conditional use of 'absolute time' is based on the regularity of the earth/sun relationship, although, being subject to minor variations, it presents in the long run a conditional 'absolute' as defined above.

As an example of a minor variation in the flow of conventional calendar time, I like to refer to a calculation made by Keven Pang, an astronomer at the US Space Agency's Jet Propulsion Laboratory in Pasadena, who found that in 12,000 BC the earthly day was one 47 thousandth of a second shorter than it is today.

Absolute Time

Other examples are the tides, changing weather patterns, ocean currents, polar ice variations – all affect the earth's rotation. The net result is that it is slowing, losing only a few thousandths of a second per century. Some geophysicists have calculated that half a billion years ago the planet travelled 20 hours to make it through the day. In another 200 million years an earth day may last 25 hours.

All this shows that the conditional use of 'absolute time', and I stress the word conditional, equating it with the flow of regular conventional calendar time is quite justifiable. However, Hawking and Einstein disagree.

Hawking writes (p. 18):

> Both Aristotle and Newton believed in absolute time. That is, they believed that one could unambiguously measure the interval of time between two events, and that this time would be the same whoever measured it, provided they used a good clock. Time was completely separate from and independent of space. This is what most people would take to be the commonsense view. However, we had to change our ideas about space and time. Although our apparently commonsense notions work well when dealing with things like apples, or planets that travel comparatively slowly, they don't work at all for things moving at or near the speed of light.

Hawking's dismissal of 'absolute time' seems to ignore the fact, shown in preceding examples, which show that all time measurements, both in practical and theoretical physics, are, without exception, based on conventional calendar time, irrespective of whether they deal with the time measurement of macroscopic or microscopic events, and irrespective of whether these events take or took place within the confines of the minuscule space of the atom, or occurred within the universal space of the cosmos. As regards measuring the speed of light, we have shown that it is unequivocally based on the conventional second and the conventional metre. Before Hawking, Einstein had dismissed the concept of 'absolute time' as outdated, describing it (B39) as **'a steady, unvarying, inexorable, universal time flow, streaming from the infinite past to the infinite future'** [emphasis supplied]. What is remarkable to observe here is that the 'regularity aspect' of time, inherent in Einstein's steady, unvarying, inexorable, universal time flow, is embodied in both Einstein's definition of absolute time, which he dismisses, and in conventional calendar time, which he accepts. He noted (B40) and I repeat: 'ALL THE CLOCKS EVER USED HAVE BEEN GEARED TO OUR SOLAR SYSTEM'.

Hawking's particular contradiction about the time concept arises from the fact that having, on p. 18 of his book, rejected the regularity aspect of 'absolute time' as defined by Einstein (above), he revives it later in the text, on p. 118, by counting the age of the universe in billions of regular conventional calendar years. This, it must be emphasised, is a method of time measurement which **only makes sense if the years in question follow each other in a steady, invariable and perfectly regular time sequence, streaming from the beginning of the universe into present times – thus equalling Einstein's definition of 'absolute time'**. This shows in other words, that Hawking rejects the regularity aspect of 'absolute time' in theory, yet uses it in practice. Other examples where Hawking routinely uses regular calendar time are his unconditional acceptance of the speed of light, **expressed in regular conventional seconds**, as well as his use of Einstein's formula $E = mc^2$, where c stands for the speed of light, equally **expressed in regular conventional seconds**. Many more examples abound where Hawking and Einstein use ordinary terrestrial calendar time to measure astronomic

Absolute Time

events in the universe and therefore inadvertently **make conditional use of what they define as 'absolute time'**. They reject this term, yet it is identical with the regular flow of calendar time as defined by Einstein above.

What nevertheless needs emphasis is that **the unconditional use** of the term 'absolute time' in physics is not advisable, since any specific time scale, as for example that of regular conventional calendar time, is bound to a particular set frame of reference, in relation to which its time system is elaborated. In the case of terrestrial calendar time the time scale is synchronised with clocks lodged on earth, which are strictly attuned to the earth/sun relationship. However, as there are other possible systems of reference in the universe, on which time relationships might be based, the general and unconditional use of the term 'absolute time' should be avoided. Instead, **all references to time should be made in relation to regular calendar time**.

7
Can Time Slow Down?

The preceding account indicates that the principle of 'regular calendar time' is of universal validity and does not allow for any irregularity in its regular flow. Most importantly, its measurements are strictly attuned to the astronomical regularity inherent in the earth/sun relationship.

The role of clocks and other timepieces
What needs stressing here is that timepieces, that is, clocks, watches and other chronometers, are only of practical use if they record a **regular time flow** strictly in step with the solar calendar. Timepieces which consistently fail to do so are useless and ought to be discarded. For reasons of this compulsive regularity, whether in ordinary life or in theoretical physics, no allowance can be made for time irregularities, specifically time slow-downs. Any irregularity which may inadvertently occur in the time flow diverting from the earth/sun relationship has to be corrected and reattuned to the **regular passage** of, let us say, Greenwich Mean Time. This proceeds just as regularly as any time sequence in other time zones of our globe, or even elsewhere in the universe, because, even though differing astronomical situations may cause the time in different time zones to differ, they can never be allowed to differ in the **regularity of their relationship**. In other

Can Time Slow Down?

words, to be of any practical use the duration of their seconds, minutes and hours must be absolutely identical and regular, irrespective of where they are located or measured, and irrespective of the 'when' from which the time was measured, either during the 'present', retroactively over billions of years, or during any intermediate period.

Accordingly, all macroscopic and microscopic events in our universe can be attuned to regular calendar time (subject to minimal corrections). This includes the speed of light, proceeding at the rate of 300,000 km per conventional second, as well as Einstein's famous formula $E = mc^2$, wherein c also denotes the speed of light expressed in conventional terrestrial seconds. Equally, the distances between earth and sun (amounting to eight light minutes), as well as the distances between earth and the fixed stars, which can amount to hundreds of light years, as well as the distances between galaxies, are all measured in terms of light years based on regular calendar time and in regard to distance are based on our common metre. Furthermore, the estimated age of the universe, approximating to fifteen billion of our calendar years, is also attuned to our regular terrestrial time standard.

Here the question must be asked: what are the consequences if time could actually slow down as postulated by Hawking and Einstein? The answer is that since time is strictly synchronised to the earth/sun relationship, a slow-down of time would have to be reflected in the earth/sun relationship and would consequently require a slowing down of the celestial movements of the earth around its own axis and in the parallel slow-down in its movement around the sun. If, as Hawking and Einstein further postulate, time can come to a total standstill (which allegedly happens when clocks attached to a light ray travel at the speed of light and stop recording any further time progress), this would require that the earth came to a

Has Hawking Erred?

total standstill in its movement around the sun. As a consequence we would experience a permanently total night on one side of the globe and a permanently lasting day on the other. Furthermore, if, as Penrose points out (P 302), mathematical equations can prove the possibility of a time reversal, then to be in accordance with it, the movement of the earth around its own axis and its rotation around the sun would require the earth to reverse its track and move in the opposite direction. Penrose writes (P392):

> All successful equations of physics are symmetrical in time. They can be used equally well in one direction in time as in the other. The future and the past seem physically to be on a completely equal footing. Newton's laws, Hamilton's equation, Maxwell's equation, Einstein's general relativity, Dirac's equation, the Schroedinger equation – all remain effectively unaltered if we reverse the direction of time . . . The whole of classical mechanics with the 'U' part of quantum mechanics, is entirely reversible in time.

What I am pointing out here reflects the true dilemma and paradox inherent in the Hawking/Einstein postulated possibility that time irregularities are endemic in the universe. Their basis is to be found in their mathematical equations, which, while theoretically sound, have totally lost any compatibility with natural events and are, rather, the result of what might be called a blinkered obsession with the infallibility of mathematics.

According to Hawking and Einstein, the slowing down of time, a main pillar of the 'special theory of relativity', can be theoretically proved by means of the Lorentz Transformation. The mathematical calculations of this are detailed in a later chapter under the heading 'the Laws of Transformation', which questions the validity of some of

Can Time Slow Down?

its findings, specifically those dealing with 'the Addition of Velocities'.

Hawking (H20) refers to several attempts made between the years 1887 and 1905 to interpret the results of the famous Michelson–Morley experiment. Most notable among them was that of the Dutch physicist Hendric Lorentz who tried to explain the results of the experiment in terms of objects contracting and clocks slowing down when moving through the ether. However, a hitherto unknown clerk named Albert Einstein, who worked in a Swiss patent office, published a paper in 1905 which proclaimed that the whole idea of an ether was unnecessary as long as one was prepared to discard the idea of absolute time.

The changes indicated by Lorentz are independent from a clock's mechanism and the structure of the measuring rods. The laws governing the slowing down of clocks and the contractions of measuring rods are defined by the 'Lorentz Transformation'. These laws are very simple, insofar as the contraction of objects results from their greater speed. Thus, a yardstick moving at the speed of light will shrink away to nothing, while a clock moving at the same speed will come to a total standstill.

What Hawking and Einstein seem to have ignored here is that light travelling at the highest possible speed shows no slowing down in its velocity, nor does it experience any shortening of distance as it ages exactly one second, when travelling the standard distance of 300,000 kilometres per second. Yet a conventional clock travelling at the speed of light is supposed to come to a total standstill, as is the age of the person carrying it, while a measuring rod is supposed to shrink to nothing! Neither Hawking nor Einstein has given us any clue as to how to explain such an obvious discrepancy. The reason is that they never seem to have tested their theory with timepieces which remain strictly aligned to the earth/sun cycle. Such a timepiece or chronometer, as it will be shown, is the conventional sundial device, although it could be any other time measuring apparatus permanently attuned to the earth/sun cycle.

Let us now scrutinise some of Hawking's and Einstein's examples of the slowing down of time by testing them against the reality of the earth/sun cycle.

Firstly, contrary to the virtual axiom of a regular time flow inherent in the solar calendar and reflected in all time measurements, Hawking and Einstein among others want us to believe that the age of persons is influenced in a sense by the behaviour of the clocks they carry. Hence the clock of a person living high above the earth's surface, (as, for example, on a water tower, see Hawking H32), will indicate that he ages more slowly than his partner below; or that a watch carried by a person travelling in space at high velocity will slow down, indicating that its wearer's age will also slow down at the same rate.

Hawking, referring to the water tower example, has remarked that this prediction was tested in 1962 on a water tower, using two very accurate clocks. One was fixed at the bottom of the water tower and the other at the top of the water tower. The bottom clock being nearer the earth's surface was found to run faster in accordance with general relativity.

In the example of the 'twin paradox', Hawking suggests that we have a pair of twins. One of the twins is destined to live at sea level while the other is going to live at the top of a mountain. When they meet again at sea level after some time it will be found that they would differ in age. The one having lived at the mountain top would now be younger compared to the other twin who had continued to live at sea level. Hawking presumes that the difference of age would however be relatively small. On the other hand, the age difference would be considerably larger if one of the twins would undertake a long journey

Can Time Slow Down?

on a space ship, travelling near the speed of light. Returning to earth after some years, he would find himself considerably younger than the twin he had left behind. This so-called twin paradox, maintains Hawking, would however remain a paradox only for those who believed in the principle of absolute time.

Einstein (B51) himself produced an example, which I will later confront with a parallel of my own. Einstein's example reads:

> In a Buck Roger realm of fantasy, it is possible to imagine some future cosmic explorer boarding an atom-propelled spaceship, ranging the void at 167,000 miles per second, and returning to earth after ten terrestrial years to find himself physically only five years older.

Adding Hawking's twin paradox example to the one above, it can now be assumed that the twin brother Einstein's explorer left behind on earth would have aged ten years, in comparison with his returning brother's five.

I am now going to show that Hawking's and Einstein's assumptions are questionable, firstly by replicating the water tower example, cited by Hawking, by using sundials as a time measuring tool instead of conventional clocks.

The idea is simple enough to be proved on paper without actually ascending a water tower. Assuming the placement of two sundials, one at the bottom and the other on top of the water-tower, it will be found that at twelve o'clock noon both dials will read almost exactly the same time, with the bottom one a minuscule part of a second in arrears, because of the distance the light has to travel between the two sundial clocks. It is important to stress, however, that this comparative reading will replicate itself exactly after every twenty-four hours without showing the slightest time discrepancy or slow-down.

Has Hawking Erred?

Going further, we now assume the presence of two sundials, one positioned on earth and exactly synchronised with the other, which is hypothetically fixed on an extended water tower, reaching half-way to the sun. It can be predicted that at exactly twelve o'clock noon earth-time (with the shadow of the earth dial pointing at twelve o'clock), the shadow of the sundial positioned half-way to the sun will indicate four minutes after twelve noon, a result of the sunlight taking eight minutes to reach the earth, and thus four minutes to reach half-way. After twenty-four hours the shadow of both sundials will again point exactly at the same notations, retaining the same time difference; and this will be repeated day after day indefinitely.

Extending the example to the relationship between the sun and Alpha Centauri (a star four light years distant), the regular day-to-day time discrepancy between the two heavenly bodies will now increase from eight minutes to four years. In neither case will any slowing down of time be observable on the sundial-measuring devices or their equivalents, even after thousands of years.

Returning to Einstein's Buck Rogers example, I now propose sending his explorer on a second trip, this time to the star Alpha Centauri. However, instead of using a spaceship, I will send him riding piggy-back on a lightbeam, travelling at the speed of light, at 300,000 kilometres per second. What ought to appear obvious is that the round trip of the lightbeam from and to earth takes about eight years, during which the lightbeam will have aged eight years (each light year equalling in duration one terrestrial calendar year). I maintain that our explorer riding on a lightbeam, ageing at the light's speed, could not have aged slower than the lightbeam, and on his return would have aged eight years exactly, like the twin brother he had left behind.

Now, according to Einstein (B49), 'A clock travelling at

Can Time Slow Down?

the speed of light (presumably a conventional mechanical clock, carried along by the explorer on his second trip) would stop completely'. Therefore, in accordance with Einstein's relativity, the explorer would also have stopped ageing. On the other hand, the lightbeam would still have increased eight years in age, because according to Einstein himself, the constant speed of light can never be subject to any time retardation, since its moving goes strictly parallel with velocity. In other words, the light covers a definite distance during a definite interval of time, which equals its ageing.

If we then assume that the explorer's progress and age, on either of his two trips, were timed by a sundial device (or indirectly by any similar chronometer lodged on earth and based on the solar calendar, his ageing would, on his second journey, coincide exactly with the ageing of the lightbeam, a total of eight years, while on his first trip (in Einstein's example), he would have aged ten years, in accordance with the sundial's measure, coinciding with terrestrial time – not ageing five years as Einstein had indicated.

Based on these considerations, the theory of the slowing down of time, advanced by Hawking and Einstein, appears to be in conflict with the true progression of natural events. As to the mathematical support of their theory, this may appear quite supportive on paper, though as we will show, it still conflicts with the reality of the natural events it attempts to illustrate. This ambiguity inherent in some mathematical paper calculations, not always matching nature's manifestations, was dealt with in more detail in chapter 3.

Meanwhile I cite two further examples with which Einstein and Hawking have attempted to prove their theory of the slowing down of time. In one, Einstein (B51) compares a radiating hydrogen atom to a kind of clock in which the time can slow down. According to Einstein, the physicist Ives, who conducted the experiment, 'Compared the light emitted by hydrogen atoms at rest, and found that the frequency of vibration of the moving atoms was reduced in exact accordance with the predictions of Einstein's equations'.

What Einstein (in accord with Ives) omits to mention is that while frequency in the two cases varied, the speed of light emitted from the same hydrogen atoms, both at rest and moving at high speed, must have remained constant and unchanged, travelling exactly at 300,000 kilometres per terrestrial second as measured by cesium clock. As the speed of light is a much more reliable permanent clock than its variable frequency, no slowing down of time can be read into this experiment.

A similar explanation can be accorded to Hawking's example (pp. 32/33), where he attempts to evaluate the 1962 watertower experiment. He points out that with the introduction of very precise navigation systems, related to satellite signals, the discrepancy in the speed of clocks at different levels above the earth's surface is now of great practical importance. He maintains that if one neglected the predictions of general relativity, the result would be a discrepancy of several miles in the calculated position of the satellites under scrutiny. As in Einstein's example of comparative frequencies, no slowing down of time does occur. What happens is that increased velocities in the frequencies allow an evaluation of the differences in frequencies at differing distances from the earth, which can aid navigation in outer space. As the speed of light remains thereby unaffected no slowing down of actual time is indicated.

A different example to repudiate the possibility that time can slow down is based on the following data supplied by Einstein. He writes (B33/61):

> In addition to its daily rotation about its axis at the rate

Can Time Slow Down?

of 1000 miles per hour and its annual revolution about the sun at a rate of 20 miles per second, the earth is also involved in a number of other less familiar gyrations. The entire solar system moving within the local star system is moving within the Milky Way at the rate of 200 miles a second; and the whole Milky Way is drifting with respect to the remote external galaxies at the rate of 100 miles a second – and all in different directions.

More remarkable still, according to Einstein, 'The astronomer's spectroscope tells him that distant constellations of stars are hurtling into limbo, away from our galaxy, at velocities ranging up to 35,000 miles per second'.

Previously we have merely dealt with humble space travellers, covering limited distances. Now we have an example where entire star systems and galaxies, involving billions of stars, many hundreds of times larger than our sun, are on the move. If we now imagine that at all these different speeds various slow-downs of time would occur, at a ratio inverse to their speed, then, considering only our own galaxy and solar system, the movements within would become snarled up in total chronological confusion. But no such slow-downs in time in the system have ever been recorded by any astronomer. On the contrary, as galaxies move away from our own galaxy (as Einstein records above), they do so with increasing speed, though with no perceptible time retardation.

If we then further assume that the light rays from all the galaxies criss-crossing the various regions of the universe do so at speeds expressed in regular terrestrial seconds (a fact fully endorsed by both Einstein and Hawking), we cannot accept that some of the regions they cross can accommodate other events which occur at a slower time sequence, despite the fact that each region comes

Has Hawking Erred?

under the same terrestrial time-standard. In other words, it appears impossible that two different calendar time sequences, one fast and one slow, can co-exist side by side in the same time-region of the universe. To assume from this that time based on conventional calendar time can slow down in some parts of the universe, and even in some cases come to a total standstill, while light travelling through the same space sector retains its regular speed and ageing in terms of terrestrial time, is an absolute anomaly. There can therefore only be one universal time scale, which, as previously exemplified, must be based on the regular flow of conventional calendar time.

In re-examining the possibility of a time/age/clock slow-down, let us reiterate the following facts. Hawking has told us that the universe is about 10 to 20 billion years old. This implies not only the use of conventional terrestrial calendar years, but also that these years follow each other in perfectly regular sequence. Further, all events occurring during these 10–20 billion years, whether involving objects in motion or at rest, must occur in unison with the criteria of terrestrial time. However, to determine the duration of such events fairly it is not possible to use conventional clocks attached to objects, because according to Hawking and Einstein, they are liable to slow-downs when travelling at high velocities, or even to stop altogether when reaching the speed of light. Such timepieces are therefore useless in measuring events of cosmic proportions. It is therefore necessary to use clocks which remain unaffected by gravitational influences when moving at great speeds. Thus only the use is applicable of timepieces or chronometers which remain strictly attuned to the solar calendar, such as sundials (as shown in the repeat of the water tower experiment) or any other chronometers which respond to light. To conclude, **it is a cast-iron rule that all human time measurements in the universe are made indirectly through a relationship to**

Can Time Slow Down?

earth-bound timepieces based on regular conventional calendar time. As a matter of fact, neither practical nor theoretical physics has ever used any time measure other than that related to regular **earth-bound** calendar time.

We have shown that all measurable events in the universe must proceed at the same regular rate, in unison with 'regular calendar time'. Therefore to synchronise extra-terrestrial events, measurements must be based on light-operated timepieces, as explained above, or based indirectly on earth-bound chronometers. Granted the use of such measurements, Hawking's space traveller of the twin paradox, as well as Einstein's Buck Rogers space explorer, not to forget my own example of an astronaut travelling at the speed of light would all have aged in perfect unison with 'regular calendar time'. And their ageing would not have differed from that of those they left behind, without regard to the number of years they might have been absent from earth or the speed they travelled at. This regular ageing process would not have differed, irrespective of whether or not the conventional clocks the explorer carried slowed down, and irrespective of whether the explorer travelled near or at the speed of light.

Einstein has pointed out (B51) that:

> The human heart is also a kind of clock. Hence according to relativity, the heartbeat of a person travelling with a velocity close to that of light would be relatively slowed, along with respiration and all other physical process. He would not notice this retardation because his watch would slow down in the same degree. But judged by a terrestrial time keeper he would 'grow old' less rapidly.

Accordingly, on reaching the speed of light, not only would the watch a person carried stop, but he himself would stop ageing. Furthermore, his entire metabolism would also come to a standstill putting him into a state of suspended animation – all of it as theoretically predicted by Einstein.

Here the question arises of what the consequences would be if the above Einsteinian principles of ageing were truly applicable throughout the universe, including our earth. Firstly, all people living on our planet with a slower heartbeat would age more slowly than those having a faster one. And if their heart stopped beating altogether, they would then hibernate in a state of suspended animation, becoming, so to speak, immortal in a biological sense. Secondly, people carrying watches out of tune with, let us say, Greenwich Mean Time, would, if their watches ran slowly, age at a slower rate than people carrying faster-running watches.

Here the objection may be raised that I have not taken note that Einstein's and Hawking's space travellers aged more slowly because they travelled at high velocities. My reply is, why then use the argument of the slowing of the heartbeat and the slowing of watches or clocks as a proof that a person has aged in real terms, when the same criteria, applied under earthly conditions, exercise no real influence on the actual ageing of persons?

As a matter of fact, the real ageing of a person (or of any other object in the universe) has nothing to do with a slow-down of the heartbeat or the slow-down of conventional watches or clocks, whether measured here on earth or far out in space. This ageing can only be assessed in synchronisation with 'regular calendar time'; all events and objects in the universe, whether organic or inorganic, age in relation to conventional human time standards, at the same regular rate of speed, irrespective of where they are located in the universe and irrespective of the velocities at which they move. **It is only under such terms that Hawking's estimate of the age of the universe and**

Can Time Slow Down?

the age of other cosmic events, as well as his estimate of the speed of light can find practical application.

If this were not so, and an actual time slow-down could happen, the paradoxical situation would arise where the time slow-down had to be accompanied by, and synchronised with, a parallel slow-down in the earth's movement around the sun. And if time were to come to a total standstill, the earth would have to stop moving altogether.

In this respect I like to mention a curiosity, claiming a standstill and a backward move of time, reputed to have happened in biblical times. (The claim was commented upon by Professor R. W. Loftin of Philadelphia in a contribution to the *Sceptical Inquirer USA*, Summer, 1991.) It was pointed out that astronauts and space scientists at Greenbelt in Maryland claimed that they had located a missing day in 1969 and even earlier. Space scientists were checking the position of planets by computer to avoid collision with satellites. The computer discovered a missing 23 hours and 20 minutes at the time the prophet Joshua made the sun stand still and an additional 40 minutes when Isaiah made the sun move backward at a sign of Hezekiah.

Let me now reiterate my point about the regularity of our basic concepts of time and space. We have shown that physicists the world over measure all physical events in the universe in units of regular calendar time, and distances in regular metric units. These regular measurements do not permit any slowing down of time to occur, nor a shortening of distances. However, in the face of such all-pervading, universally applicable regularities, Einsteinian relativity maintains that clocks attached to moving objects which travel at high velocities will slow down, while measuring rods carried along will shrink, thereby influencing the regularity at which the objects move. Following the same principle, clocks, on reaching the speed of light, are supposed to come to a total standstill, while measuring rods are assumed to shrink to nothing.

Yet we know by indirect measuring, that, for example, light travels at 300,000 km per second, and does so in a perfectly regular manner. If therefore a clock and a measuring rod were attached to a light beam to measure its speed, the clock would record that the light had not aged even the tiniest fraction of a second, while the measuring rod would record that the light had not moved even a millimetre. Such reasoning shows that moving clocks and moving measuring rods can in no way influence the real measurements of time and distances of objects in motion; and that the assumption that they can do so is pure fiction, having no basis in physical reality.

8
The Concept of Space

An examination of the space concept reveals several twists in its possible interpretation, some of them being of a semantic nature. Most important is the necessity for a strict differentiation between voluminal and linear space and their relation to time.

Reading through the various contributions of Hawking and Einstein on the subject of 'space' one finds much ambiguity in their use of spatial facets. Often their distinction between linear and voluminal space becomes blurred, particularly in their treatment of the vexing subject of space-time, and even more so in respect of the alleged curvature of space. (Both are dealt with in separate chapters below.) For instance, according to Hawking (H23), time cannot be considered as something independent, or separate from space, but forms, as he asserts, an object termed space-time. He states (H33):

> In the general theory of relativity space and time are dynamic quantities: when a body moves, and a force acts, it affects the curvature of space and time – and in turn the structure of space-time affects the way in which bodies move and forces act.

Hawking does not tell us to which type of space the

curvature refers – whether it relates to one-dimensional linear space or to three-dimensional voluminal space.

Dictionaries generally describe 'space' as:
1. The boundless expanse in which objects exist and move.
2. An area or surface of objects.
3. An interval between two points or objects.
4. An interval of space expressed in terms of time.

These varying descriptions, although pointing at the multifarious nature of space, are inadequate for our purpose and need further elaboration. I believe that the following definitions can help to clarify the issues involved.

Firstly, there is one-dimensional linear space

which separates two points or objects. It determines the distance between static bodies and the speed or velocity of moving ones. The latter involves the use of time, in addition to linear distance measurement. On the other hand, plain distance measurement between static points or objects such as a line drawn on paper or a railway track laid across country, does not involve any time relationships.

Secondly, there is two-dimensional area or surface space

which is determined and measured by the two coordinates, length and width. This two-dimensional space concept is purely abstract, insofar as it expresses a mere attributive property (the surface area), with which all material bodies are endowed irrespective of their shape or size. Surface space can be identified by its colour or shape or by any other surface particularity which covers or envelops objects, and this can be done in a strictly non-material, attributive way. The same principle applies to the surface of a fluid, though with slight differences in perception. Furthermore, surface or area space is a permanent feature of all material objects, whether solid, fluid or gaseous, none of these features having any specific time relationship. As an example I can point to the surface of my table on which I write this sentence. It has no intrinsic time relationship, with one exception – it ages, together with the table, though only theoretically, having no material existence.

An example follows of how the two-dimensional space concept has been misrepresented in a truly fantastic way by Hawking. It is accompanied by the explanatory illustration of a two-dimensional animal, taken from his chapter 'The Unification of Physics' (H164).

According to Hawking, in the creation of complicated creatures like humans it appears that two space dimensions are not sufficient to sustain them. He supposes that if two-dimensional creatures were to live on a planet that had merely one dimension, in order to pass one another such animals would have to scramble over each other's back. Furthermore, a two-dimensional animal, when it came to eat, would hardly be able to digest any food. Instead it would have to void the food in the same form in which it had swallowed it, as the digestive passage would cut the animal in two. Therefore, any two-dimensional creature would not only be unable to digest food, but would have fallen apart even before attempting to eat. Similarly, the presence of blood circulation in a two-dimensional being would present a total anomaly.

Such an example presents us with one of Hawking's most remarkable aberrations. To imagine a two-dimensional being, it must have besides length and height also thickness (which immediately would transform it into a three-dimensional being) even to eat or to scramble over another such being. Even to think of a two dimensional being is a total anomaly, because the term two-dimen-

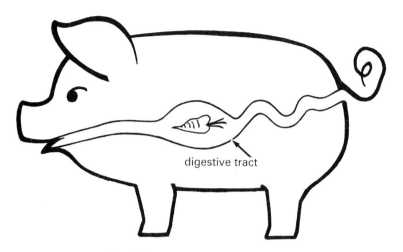

Fig. 2: The two-dimensional animal.
(Hawking's example is a dog).

sional is only an attributive expression, describing a purely non-material, hypothetical situation. Also, speaking of a one-dimensional world is a total anachronism, because even in the most vivid imagination such a world cannot have any material existence whatsoever.

One must therefore not be surprised to find Hawking, in common with Einstein (as it will be shown later in the text), endowing the abstractions which denote universal voluminal space, as well as time, with the capacity of bending, warping, curving and even flattening (like the above animal), though they have no material existence whatsoever.

Thirdly, there is universal voluminal space
possibly the most important type to be referred to here. It can be described as an all-pervading, non-material, abstract entity which envelops the entire universe and which is static, as well as completely indifferent to the

The Concept of Space

passage of time. It actually fits Hawking's description of 'absolute space', i.e. 'a shapeless, steady, hypothetical void, which can be conceived as empty space and which is static, timeless and of infinite dimension.'

Hawking dismisses 'absolute space' as described by him above as obsolete and outdated. And this goes hand in hand with his dismissal of 'absolute time' as defined by Einstein (see chapter 5 on absolute time). The reason is, of course, that an acceptance of the principle of 'absolute space' and of 'absolute time' would undermine the fundamental premise on which the theory of relativity is built and which demands that 'space' (i.e. 'universal voluminal space') is a malleable substance and that 'time' is an unsteady and variable phenomenon.

To pursue our comments on 'universal voluminal space' further, we can point out that it houses all the three-dimensional voluminal objects and substances extant in the universe and incorporates all one-dimensional configurations, whether time-bound or independent of time, as well as all two-dimensional configurations (i.e. flat surfaces or surfaces enveloping solid objects), which, being merely attributive abstractions, have no time linkage by their very nature as non-material objects.

A remarkable example where a three-dimensional space object (a piece of string) is given a material existence, though it is supposedly a strictly one-dimensional phenomenon, is advanced by Hawking (H159), in commenting on the so-called 'string theories'. Hawking explains that in string theories the essential objects are not particles occupying specific positions in space, but are things that have only one dimension, namely length, thereby resembling infinitely thin pieces of string. Strings having ends are called open strings, while strings forming closed loops are called closed strings. In contrast, a specific particle occupies a certain point in space at any particular moment of time, so that its history is traceable as a line in space-time, the so called 'world line'. A string, on the other hand, at any specific moment of time occupies a line in space which leads to its history in space-time forming a two-dimensional surface called 'world sheet'.

The flaw in this example is the assertion that the strings (intended to represent an alternative to conventional particles) have only length and no other dimension. Now it is quite logical that even the infinitesimally thinnest string, whatever its structure, must have a material existence and be a three-dimensional voluminal space object. Alternatively a one-dimensional space configuration is purely descriptive and must always remain abstract and non-material. If, in defiance of this principle, it is asserted that the strings have only length, i.e. are one-dimensional, it is clear that they cannot have any material existence whatsoever and in fact cannot exist in a one-dimensional form. Strings in this context embody the same contradiction as Hawking's two-dimensional dog. However, Hawking not only passes over this conceptual error but conceals it within the enigmatic phraseology of space-time, which is intended to rationalise it.

Consideration must now be given to the question of whether universal voluminal space – the space that contains the universe – is basically empty.

At one time it was believed that the entire universe, including what we term 'empty space', was filled with a hypothetical substance called ether, the medium in which electromagnetic waves, etc., were assumed to be transmitted through space. For example, ether was postulated by Maxwell to establish the laws of electromagnetism. The concept of ether has since been discarded, but to date no rational analysis has been put forward to explain how light waves and other similar media (including energy) can travel through supposedly empty space without some medium to support them. It must therefore be asked whether an alternative to ether has yet to be postulated,

The Concept of Space

to act as an energy transmitter, filling the entire universe and therefore negating the concept of 'empty space'. However, even were there no empty space, our three space definitions proposed above would still apply, as **they are independent of the concept of 'empty space'**. On the other hand, Einsteinian relativity maintains that the acceptance of space-time does away with the need for ether or any substitute for it.

In a more detailed analysis, universal voluminal space reveals itself as a purely hypothetical, intangible entity, timeless and, like surface space, with no material existence of its own. The all-embracing, timeless and spatially limitless void, which we conceive as universal space, houses everything from the biggest galaxies down to the minutest particles of matter. Also included in it is what Einstein describes as a universal gravitational field. Again, all these objects, whether at rest or in motion, occupy part of the volume of universal space, in their own three-dimensional presence.

This definition does not exclude the possibility that universal space may not be entirely empty, but contains, for example, the universal gravitational field postulated by Einstein. However, even if universal voluminal space were entirely full, it still implies the presence of the hypothetical entity of universal space in the first place, so as to accommodate the gravitational field in the second place. The principle involved may be compared with the space in a presumably empty bottle filled with air, except that universal voluminal space must be imagined as occupying a vessel of infinite dimensions. In other words, the hypothetical abstraction 'space' (in the sense of voluminal space) must be present whether it is full or empty.

Einstein himself has pointed to this fact in a footnote on p. 155 of his book *Relativity*, (1988), in the chapter 'Relativity and the Problem of Space'. He writes, 'If we consider that which fills space (i.e. the field) to be removed, there still remains the metric space.' However, in the very same paragraph, he points out that: 'There is no empty space, i.e. a space without field.'

Penrose, who like Einstein and Hawking unreservedly believes in space-time, comments thus on 'empty space' (P240): 'One implication of Maxwell's equation was that electric and magnetic fields would indeed 'push' each other along through **empty space**' [emphasis supplied]. Writing about mass and energy, he argues further (P285) that:

> . . . if the latter is to be located at all it must be **in this flat empty space** – a region completely free of matter or fields of any kind. In these curious circumstances, our 'quantity of matter' is now **either there, in the emptiest of empty regions, or is nowhere at all** [emphasis supplied].

On another occasion, Penrose (P436–7) speaks of an astronaut travelling in **'empty space'**, and (P269) states that, '. . . rays of light (world lines of photons) in **empty space** are also geodesics'. One ought to assume therefore, that in contrast to Einstein, Penrose accepts the reality of 'empty space', which, to have any theoretical standing, must equal 'absolute space' as defined (though rejected) by Hawking.

All this may sound rather involuted. Nevertheless, I believe that the problem of space can be presented in such a way that even the layman can understand its implications. Fundamentally, we can say that 'space' is a descriptive term which, in the course of the development of ordinary language, became incorporated into our vocabulary, allowing us to rationalise it. As already pointed out, the term 'space' itself, like other expressions of multifarious meaning, lends itself to different interpre-

The Concept of Space

tations, describing a) linear space-distance, b) surface-space, c) volume.

To give a practical example, let us consider an unfurnished, empty rectangular room. The space, or distance between the walls, as well as that between floor and ceiling, can be measured linearly. The surface of the four walls, as well as those of floor and ceiling, can easily be identified as surface spaces, while the hypothetically empty space occupying the room is the room's volume. The fact that the room will be normally filled with air is of no significance.

We assume now that this room is part of a house separated from another house by a measurable linear space-distance. We also find that the first house has a large backyard and a front garden, while the second house is situated in front of a large playing field. There is no difficulty in describing the backyard, front garden and playing field and other neighbouring open floor spaces as surface spaces, all measurable in two dimensions. Furthermore, the voluminal spaces above and between the houses, as well as those above the backyard, front garden, playing field and any other neighbouring open space, stretch upwards into limitless distance, reminding us of the saying 'the sky is the limit'. Readers will also be acquainted with the term 'stratosphere' which typifies the space above the earth's atmosphere. All the voluminal spaces mentioned above are again part of the three-dimensional universal voluminal space which occupies the universe. Scientifically considered, universal voluminal space is a purely abstract entity which has no material existence.

If we now also consider that there are material objects located in the open spaces around the houses, such as trees, fences, telegraph poles, stones, etc., there is no difficulty in identifying them as three-dimensional space objects, each of them occupying part of the voluminal empty space around the houses. What is further obvious

Has Hawking Erred?

is that all the above special configurations, when applied to the practical needs of everyday life, present no serious conceptual difficulties to persons of average intelligence.

From the above details about space it is further obvious that universal voluminal space, whether empty or occupied, is independent of the passage of time. On the other hand, any three-dimensional objects when in motion can be time-bound, insofar as their velocities are measurable in relation to regular conventional calendar time.

Let us now assume that I pick up a handy stone and throw it into the open space above me, as far as I can. The stone will describe an arc of 40 metres, reaching the ground again after a flight of, let us say, five seconds. What is obvious here is that the stone will not have lost its three-dimensional qualities while moving through the space above and will land unchanged in volume.

Based on this stone example, I venture to suggest (as further detailed in later chapters), that the same unchanged qualities will prevail if I now extend the example to an aeroplane, a spaceship, the movement of our earth around the sun and around its own axis, the movement of entire galaxies, and the movement of a humble worm, crawling along the ground. None of these objects will be affected in their three dimensionality, whether moving at snail's pace or near the speed of light; although in the latter case (that of high velocity), the shape and mass of the moving objects may be subject to change, though not its three-dimensional quality. To assert, however, that moving three-dimensional material objects can gain an additional space dimension is a total illusion. Further, the three-dimensional voluminal space through which all these objects travel will not be affected by the object's passage. Empty voluminal space is a purely hypothetical non-material medium which plays only a passive part in the process and remains totally unaffected by objects passing through it, whatever their speed or

The Concept of Space

composition. It is also worth mentioning here that on one occasion in his earlier writings (*Relativity*, 1916 edition, incorporated in 1988 into the book's 15th edition), Einstein, in a passage on the subject of 'Space in Classical Physics', showed himself to be highly mystified by the space concept. Referring to 'The System of Coordinates' (an earlier chapter in his book), he wrote:

> We entirely shun the vague word 'space', of which, we must honestly acknowledge, we cannot form the slightest conception, and we replace it by 'motion relative to a practically rigid body of reference'.

In his later writings 'space' seemed to have shed much of its mystery, when he presented a description of space examples (see B57–8), which more or less match the descriptions of the threefold nature of 'space' proposed in this book.

For a one-dimensional (linear) space continuum he uses a ruler and a railway track; for a two-dimensional space continuum he uses the sea's surface; for a three-dimensional space continuum he submits the flight of an aeroplane, writing:

> An aeroplane pilot guides his plane through a three-dimensional space continuum, and this constitutes space as we perceive it. In other words, the space of our world is a three-dimensional continuum.

All this is quite acceptable, except that he fails to mention that this space continuum has no discoverable bounds.

Einstein also tried to make us visualise the non-material nature of space (i.e. of universal voluminal space) by writing (B14) 'Space has no objective reality except as an order or arrangement of objects we perceive in it.' This is a definition which, when everything is considered, implies that space is an abstract, non-material entity matching our own view on space. What is remarkable here is that Einstein's view on space, which he shares with Hawking, is identical with the two scientists' definition of 'absolute space', a concept which both have rejected as obsolete and outdated.

Absolute space

As already indicated, Hawking described 'absolute space' (which he rejects) as 'a shapeless, steady, hypothetical void, which can be conceived as empty space and which is static, timeless and of infinite dimension.' Special reference must be made here to Newton (see B34), who presumed that space itself might serve as a fixed frame of reference to which the wheeling of the stars and galaxies could be related in terms of absolute motion. Newton regarded space as a physical reality, stationary and immovable; while he could not support this conviction by any scientific argument, he nevertheless clung to it on theological grounds, for to Newton space represented the divine omnipresence of God in nature.

Einstein (B38) rejected the Newtonian idea of space (i.e. of 'absolute space'), that is, of a fixed system or framework, absolutely at rest, within which it is possible to distinguish absolute from relative motion, saying that 'in space there are no directions nor boundaries.'

We thus find ourselves with two differing descriptions of 'absolute space', both rejected by Hawking and Einstein as outdated:

a) the Newtonian one, resting on a materially based, perceptible frame of reference:

b) the one put forward by Hawking above, in which space does not provide a Newtonian-type fixed frame, but which, to repeat Hawking, is definable as a 'shapeless, steady, hypothetical void, which can be conceived of as

The Concept of Space

empty space and which is static, **timeless** and of infinite dimension'. I have emphasized the word 'timeless', because it is the separateness of time and space contained in the above definition of 'absolute space', which forms the essence of their rejection of absolute space as defined by Hawking. This rests on the fact that Einsteinian relativity postulates an intricate relationship between time and space, treating time as an extra space dimension resulting in the concepts of space-time and space-curvature. Einstein writes (B62):

> An understanding of the space-time continuum is requisite to a comprehension of the General Theory of Relativity and of what it says about **gravitation, the unseen force that holds the universe together** and determining its shape and size [emphasis supplied].

The curious result of all this is that space-time and space-curvature in association with Einstein's universal gravitational field can be construed as a new fixed frame of reference for astronomical calculations, and thus be considered as a veiled retreat into a Newtonian-like, though not identical, 'absolute space'. This is indicated in the arguments about gravity in my next chapter, because by associating space-time and space-curvature with his universal gravitational field, Einstein did create a new universal absolute to which, according to him (and in retrospect to Newton), the wheeling of the stars and galaxies can be mathematically related.

Before going further, I have to point out that if it can be shown that 'space' and 'time' are two entirely different phenomena which cannot intermingle (and I will try to show this in the later chapter on 'Space Time'), then the above definition of 'absolute space' by Hawking (which both Einstein and Hawking dismiss as outdated) must be considered as a scientifically sound, acceptable proposition, which conforms with natural events.

9
Empty Space and Gravity

While there can be little doubt that gravity is the unseen force which holds the universe together and regulates the movements of the biggest galaxies as well as those of the tiniest subatomic particles, as they proceed within the timeless bounds of universal voluminal space, it does not conform with Einstein's concept of 'space-time'.

In his chapter on elementary particles, p. 70, Hawking describes gravity as the weakest of the four forces, pointing to its two specific properties: the capacity to act over very long distances; and the possession of a permanent attraction. Though weak when exercised by tiny particles, gravity can become a considerable force when these particles exercise their combined attraction, as in the case of sun and earth or any other heavenly body.

This conforms with Einstein's views (*Relativity*, 1988, p. 64) where he writes:

> The law governing the properties of the gravitational field in space must be a perfectly definite one, in order correctly to represent the diminution of gravitational action with the distance from operative bodies. It is something like this: The body (e.g. the earth) produces a field in its immediate neighbourhood directly; the intensity and direction of the field at points further

removed from the body are thence determined by the law which governs the properties of space of the gravitational fields themselves.

And on p. 65, Einstein refers to the law that 'The gravitational mass of a body is equal to its inertial mass.'
A more comprehensive account of the 'gravitational field' pointing at its universal application is contained in his chapter 'Relativity and the Problem of Space' (*Relativity*, 1988, p. 155):

In order to be able to describe at all that which fills up space and is dependent on the co-ordinates, space-time or the inertial system with its metrical properties must be thought of at once as existing, for otherwise the description of **'that which fills up space'** would have no meaning. On the basis of the general theory of relativity, on the other hand, space as opposed to 'what fills space', which is dependent on the co-ordinates, has no separate existence. Thus a pure gravitational field might have been described in terms of the 'gik' (as functions of the co-ordinates), by solution of the gravitational equations. If we imagine the gravitational field, i.e. the functions 'gik', to be removed, there does not remain a space of the type (one), but absolutely 'nothing', and also no 'topological space'. For the functions 'gik' describe not only the field, but at the same time also the topological and metrical structural properties of the manifold. A space of the type (1), judged from the standpoint of the general theory of relativity, is not a space without field, but a special case of the 'gik' field, for which – for the co-ordinate system used, which in itself has no objective significance – the functions 'gik' have values that do not depend on the co-ordinates. **There is no such thing as an empty space**, i.e. a space without field. Space-time does not claim

Empty Space and Gravity

existence on its own, but only as a structural quality of the field [emphasis supplied].

A further contribution to this subject is Einstein's remark (B97) that 'relativity reduced gravitational force to a geometrical particularity of the space-time continuum.'

I believe that the above account indicates the postulation by Einstein of a universal gravitational field, possessing geometrical attributes, which, all things considered, can be interpreted as the equivalent of a new Newtonian type space-absolute.

Penrose refers to another aspect of gravity (P157–60), when he writes about the physical effects of quantum gravity: 'They may occur at the absurdly tiny distance known as 'Planck Length' – which is 10^{-34} metres, being 100,000,000,000,000,000,000 times smaller than the size of the smallest subatomic particle.'

What seems obvious from this description is that the absurdly tiny distance referred to above must be empty space, separating the elusive particles called gravitons from each other. But if this were the case, then Einstein's assumption that there is no empty space in the universe would falter, since one could conjecture that if there is no empty space between gravitons, these would form a coherent soup filling entire universal space, through which no other particles could pass – thus putting the very existence of the universe in question.

10
Space-time Analysed

For thousands of years, ever since language bestowed any meaning on these concepts, space (i.e. universal voluminal space) and time have been treated as two distinct natural phenomena, each of them serving a different function. This distinction of function is shared by other natural phenomena such as temperature, sound, weight, and so forth. Just as the functions of temperature and sound do not intermingle, so neither do the functions of space and time, or those of time and temperature: all this in spite of the fact that all events occurring within the realms of universal voluminal space are time bound and temperature bound, while space in this scenario serves as a passive medium. Still, I do not deny that time (in its duration measured by clocks) has definite space relationships to linear space – but time duration, as it will be shown, can never substitute for linear space distance. Furthermore, time and space must always remain different phenomena, and time can never be assumed to function as an extra space dimension, neither in linear nor voluminal space.

Before going into details about the intricate subject of space-time I find it necessary to recall previous definitions of time and space:

Space-time Analysed

Time

In its everyday practical use, 'time' (like metric measure), is no more and no less than a comparative means of measure, which has no corporeal existence of its own. It is a specific human invention, tailored to compare and measure the intervals between passing events, the speed of moving bodies, the ages of objects, in motion or at rest, the age of growing organisms and, in a global application, the age or duration of the universe within which all these events occur. Furthermore, all these occurrences are attuned to our conventional calendar time and its subdivisions. They are based on and regulated by the astronomical relations between earth and sun and have been used ever since time measurements attained any degree of accuracy. I recall here again Einstein's remark (B40): **'All clocks ever used have been geared to our solar system'** [emphasis supplied].

Space

In considering the different functions of time and space one should assume as axiomatic that time measurements express duration and linear space measurements distance, and that voluminal space is strictly three-dimensional. This should further imply that linear distance cannot be measured by clocks but only by metric units of length. If one speaks therefore (and this has become common usage) of a) a 'light second' or b) a 'light year', implying a linear distance, this is merely a shorthand expression which actually means a) the linear distance light travels during one ordinary second and b) the linear distance light travels within one calendar year's duration. As to voluminal space, its three-dimensional quality is absolute and cannot be amplified by the addition of further space dimensions. This applies equally to universal space and to all the space objects (all of which are three-dimensional), which it houses, none of which can accommodate any

Has Hawking Erred?

extra space dimensions, just as a triangle can not accommodate more than three sides.

Curiously though, this cast-iron rule has been breached by Einstein's Special Theory of Relativity, which postulates the existence of the concepts of space-time and space-curvature. My aim is to show that Einstein's postulates are not in accord with natural events, but rest specifically on a turn of phrase involving a wrong interpretation of the term 'space', with the substitution of linear space-distance by time distance. The result has been the conversion of time into an extra space dimension. The distinction between linear space and voluminal space is not strictly adhered to in the writings of Einstein and Hawking. This laxity leads to an intermingling of these two otherwise distinct space phenomena, thereby causing a great deal of conceptual confusion.

Space-time in linear space

The space-time examples given by Einstein and Hawking derive from Euclidean geometry, which in itself is merely a paper exercise; moreover, the examples point to the possibility that the anomaly of an extra space dimension may have arisen from the particular semantic interpretation to which the term 'space' lends itself. One can habitually express 'time' in terms of space, stemming from phrases such as: **a train covers a distance of five kilometres in the 'space' of five minutes;** or light covers a distance of 300,000 kilometres **in the 'space' of one second**. In these and similar examples, 'space' in the sense of linear space (although basically a mere figure of speech, denoting distance), seems identifiable with time (denoting duration).

That Einstein used this particular semantic interpretation when formulating his ideas on time and space is evident in the following passage from his book *Relativity* (1988, p. 56–7). Referring to the discovery by Minkowski,

Space-time Analysed

his mathematic teacher, that time and space are one, Einstein writes: '**Pure space distance** of two events with respect to K, (K being a first coordinate system) **results in time distance** of the same event in respect to K', K' being the second coordinate system)' [emphasis supplied].

As to Hawking, he elucidates (H23) that to fix the position of a specific point in space we define it by the three co-ordinates, length, width and height. For example, a point within the voluminal space of a room might be situated three metres distant from one wall and five metres distant from another, while its height from the floor measures three metres. In another example (H24), Hawking suggests that one may measure the position of a specific point in the London area in miles north and west of Trafalgar Square; or alternatively, miles north-east and north-west of Trafalgar Square. Equally so, using relativity, one could use a new time co-ordinate, which was the old time (in seconds), plus the distance (in light seconds), north of Trafalgar Square.

In the foregoing we have an example where Hawking expresses a linear space distance in terms of time, elevating time to an extra space dimension. He subsequently concludes: 'It is often helpful to think of the four co-ordinates of an event as specifying its position in **a four-dimensional space called space-time.**' [My emphasis].

It would thus appear as if time measurement expressing duration could function as an alternative to metrical measurement expressing distance, thereby elevating time to an extra space dimension. In other words, it appears as if metrical space-distance could be substituted by time-distance (as in the example 'in the space of five minutes'), or even be transformed into time-distance or vice versa. It is possible that the semantic principle expressed in this context, although applicable here only to linear space, may have induced Einstein to generalise (the more so as the German expression 'Zeitraum' is more explicit), and along with other considerations to add time as a fourth space dimension to three-dimensional voluminal space.

Hawking also writes (p. 23): 'In the theory of relativity, **we now define distance in terms of time and the speed of light'** [emphasis supplied]. He also refers to the **light-second** as a convenient new unit of length measurement, and writes: **'This is simply defined as the distance the light travels in one second.'** [Emphasis supplied].

Hawking further points out that as space is not completely separate from time, they unite to form the concept of space-time. Just as there is no particular differentiation in relativity between time and space co-ordinates, so there is no real differentiation between two space co-ordinates.

My argument is that Hawking's conclusions drawn from the above account and other similar examples are highly questionable. Firstly, the fixed standard distance of a metre (although now expressed in fractions of a light second) was in use long before anybody thought of using light as a comparative measuring rod. Furthermore using the awkward fraction of a light second for the measurement of distances in everyday life seems quite cumbersome and is indeed impractical. On the other hand, the use of the light-second, or the light-year, for measuring astronomical distances is quite practicable.

Yet even so, it is quite illusory to assume that the velocity of light denoting distance can serve as a surrogate for a clock denoting duration. The reason is that a light-second or a light-year is only casually related to distance, because what it really denotes is duration. To define these terms correctly, one has to say that **a light-second denotes the distance light travels in one second's duration and a light-year denotes the distance the light travels during one year**. What is most important to stress here is that the second and the year in question equal in duration our conventional second and our conventional calendar year

– as a matter of fact, they are based on it. Hence, during the same one second's duration in which the light covers a space distance of 300,000 kilometres, an athlete may cover a space distance of ten metres on a running track, the light-second being exactly the same as any ordinary second. Or while over a light-year light travels a distance of 365 × 24 × 36000 × 300,000 kilometres; the duration of time involved equals exactly that of a conventional calendar year. The stem of a tree may grow merely ten centimetres taller during the same light- or calendar, year, the light-year equalling exactly, and being identical, with any calendar year.

In the statement that light covers a distance of 300,000 kilometres in the 'space' of one conventional second, which is of the same duration as a light-second, it is clear that the term 'space' stands here for 'duration', measurable on any conventional clock or stop-watch.

To illuminate the principle of space-time, Einstein provides the graphic example of a railway journey, shown below.

The space-time diagram of Einstein (B59)
A train leaves New York at 6 a.m. (Eastern Standard Time), reaching Chicago at 1 p.m. (Central Standard Time) the next day. The intermediate stations are marked with the times noted vertically and the space-distances horizontally. According to Einstein the diagram illustrates a space-time continuum, elevating time to an extra space dimension.

Einstein explains:

> To describe any physical event involving motion, it is not enough simply to indicate position in space. It is necessary to state also how position changes in time. Thus to give an accurate picture of the operation of the

Has Hawking Erred?

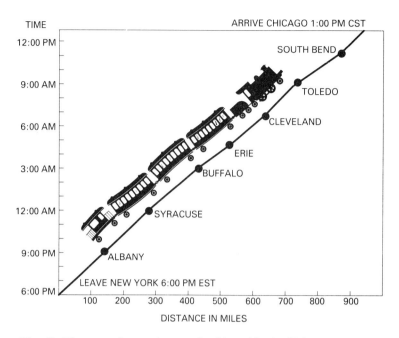

Fig. 3: The westbound run of a New York–Chicago express pictured in a two-dimensional space-time continuum

New York–Chicago express, one must mention not only that it goes from New York to Albany to Syracuse to Cleveland to Toledo to Chicago, but also the times at which it touches each of those points. This can be done either by means of a time table or a visual chart. If the miles between New York and Chicago are plotted horizontally on a piece of ruled paper and the hours and minutes are plotted vertically, then a diagonal line properly drawn across the page illustrates the progress of the train in a two dimensional space-time continuum. This type of graphic representation is familiar to most newspaper readers: **a stock market chart, for example, pictures financial events in a two dimensional dollar-time continuum** [emphasis supplied].

Space-time Analysed

On closer examination this example reveals various flaws. For instance, the train is bound to stop at all the stations listed for five to ten minutes or even longer. This means that during these stopping times, totalling perhaps one hour, the clock, or clocks, recording the progress of the train will keep on ticking, even though the train is at a standstill, covering no distance in terms of linear space at all during the intervals. This shows that Einstein's train example, which is supposed to show continuous motion (illustrating a continuum) in terms of linear space, synchronised with time-duration, and thereby illustrating space-time (a unified concept), is misconceived. What it shows rather is that while the space-distance involved as related to time is discontinuous and studded with interruptions, the related time duration shown on a clock is continuous and uninterrupted. In other words space-distance here does not match time duration: the assumed space-continuum differs from the time-continuum.

Thus, contrary to assumptions, it shows that the New York – Chicago journey, with the train covering 1100 miles in the time-space of 19 hours, does not synchronise with the progress of time during the stoppages. Hence, although Hawking's original reference to linear space-time (H23), where he writes: 'We must accept that time is not completely separate from and independent of space, but is combined with it to form an object called space-time', seems to indicate that time, superficially seen, can play the part of an extra space dimension, closer examination reveals that this is not the case.

Einstein's train example makes this quite obvious, so that we can now add that the space (or time-space) in question does not directly relate to the linear distance the New York–Chicago express travels, but to the linear distance (or space) covered by the dial of the Chicago station clock, which indicates the distance the dial moves along its circular face.

Has Hawking Erred?

The time-space, or distance traversed by the dial on the face of any conventional clock would, however, shrink to nil, if a digital or quartz clock were used. Furthermore, these time measurements in terms of space registered on the face of a clock, whether recorded circularly or digitally, occur quite independently of the linear distance (or space) the train travels during the time in question, or the time-space measured in terms of linear space on the clock's face, traversed by its dials. As the station clocks of New York and Chicago are doubtless synchronised, the clocks would also record at the same time the increase in age of all the passengers travelling on the train, which would exactly parallel the age increase of the train itself. What is also worth noting is that the age increase of the train and that of its passengers has nothing to do with the space or distance the train travels, as the age increase would remain the same whether it was measured in relation to a stationary train or to one in motion.

These train examples endorse previous conclusions that time and space belong to different categories of phenomena, just as do time, sound, temperature and weight. It can therefore also be contended that time, being a man-made comparative measuring device, recorded by clocks or other timepieces, cannot have any direct bearing on the space factor.

'Space' is an inherent part of nature, while time, or a clock recording it, is a man-made tool arbitrarily aligned to the sun/earth relationship. Just as a metre-rod lacks the inherent potency to regulate the distances between objects or to influence their velocities, and a thermometer lacks the capacity to influence temperature or weather, so does a clock, our time-measuring tool, lack the capacity to influence or determine events happening in space.

Based on such considerations, Einstein's assertion (B60) that there exists an equivalence of space and time can no

Space-time Analysed

longer be upheld; nor can we support Einstein, when he says (B60):

> It must not be thought that the space-time continuum is simply a mathematical construction. The world is a space-time continuum; all reality exists both in space and time, and the two are indivisible. **All measurements of time are really measurements in space, and conversely measurements in space depend on measurements of time** [emphasis supplied].

The dictum that measurements in space are really measurements of time is also contradicted by the following examples. In accordance with accepted theory, the sun's light travelling at a velocity of 300,000 kilometres per second will take eight minutes to reach our earth. To show further the independence of time-measurement (i.e. duration) from space-measurement (i.e. measurement of distance or volume), let us consider what happens within the space (i.e. duration) of the same eight minutes in different parts of the universe. In any region of universal space, light will travel a distance of $8 \times 60 \times 300,000$ kilometres. During the same eight minutes a snail will somewhere on earth crawl a distance of about two metres. A solid rock will merely increase eight minutes in age without showing any measurable space relationship. Equally, any human or animal, whether alive or dead, will increase eight minutes in age, again without any measurable relationship to space. And finally, the entire universe will age exactly eight minutes in perfect synchronisation with my own wristwatch or any other timepiece wherever located on earth. This also means that my wristwatch will indirectly record at the same time the duration of all the events happening in the universe, including the movement of the galaxies and the ages of all objects whether in motion or at rest, with some of these showing, and others not showing, any measurable space-relations. Furthermore, all these events occurring in the universe will take place within the expanse of universal voluminal space, which is a permanent and static void, and which is completely indifferent to the time-measurement registered on my watch or any other timepiece wherever located. In other words, my wristwatch or any other timepiece, all being man-made tools for time-measurement, can on no account influence events in the universe, whether in relation to objects in motion or objects at rest. The same principle is inherent in a thermometer. Being a mere tool for temperature measurement, a thermometer cannot exercise any influence whatsoever on temperature or the weather.

11
Time as an Extra Space-dimension

If we elevate 'time' to the status of an extra space dimension as Einstein and Hawking would like us to do and as Einstein tries to achieve in his New York–Chicago express train diagram, we can do the same thing with weight, temperature and sound, adding them as extra dimensions to space.

How this can be accomplished is indicated by Einstein himself. Having (in B58) graphically illustrated the progress of the New York–Chicago express in a two-dimensional space-time continuum, he then writes:

> This type of graphic representation is familiar to most newspaper readers; **a stock market chart, for example, pictures financial events in a two dimensional dollar-time continuum** [emphasis supplied].

This example shows that Einstein elevates time to an extra dollar dimension, although how time, recorded on a clock mounted on the wall of the New York Stock Exchange, can influence stock market quotations, is debatable. However, to compare the two-dimensional dollar-time continuum with a two dimensional space-time continuum is a rather fantastic feat, but no more fantastic than the mixing up of the two entirely incompatible natural

Has Hawking Erred?

phenomena of space and time and fusing them together into the new concept of space-time.

This incongruity becomes even more obvious when we consider Einstein's and Hawking's dictum that in a reverse process to time being a dimension of space, space can become a dimension of time. As the same principle must apply to the dollar-time relationship, we can now say that a dollar can become a dimension of time. Not only that, but having shown that the dollar can function as a time-dimension, while time can function as an extra space-dimension, it must (using Einstein's reasoning) also be theoretically possible to consider the dollar as an extra space-dimension.

However one may look at such two-dimensional manipulations, I maintain that in using Einstein's graphical illustration of a two-dimensional space-time continuum, it is perfectly legitimate to use the same principle of graphic illustration to elevate the natural phenomena of weight, temperature and sound to the status of extra space-dimensions. This is illustrated below.

The practical unit of work done by a force of one kilogram of weight over a distance of one metre equals one metre kilogram (mkg); that of 2 kg of weight over a distance of two metres will amount to 4 mkg; 3 to 3 will give 9 mkg; and 4 to 4 will give 16 mkg. The weight in kg is marked on the vertical line and the distances the weight moves are marked in metres of linear space on the horizontal line. The diagonal line indicates the respective points corresponding to weight and distance (space). Based on Einstein/Hawking reasoning we can speak here of **a space-weight continuum, elevating weight to the status of an extra space-dimension**.

A siren at point A sounds an alarm to the police posts B, C, D and E, which are marked on the diagonal line in distances 400 yards apart; the sound travelling at about 400 yards per second (the actual figure being 373 yards)

78

Time as an Extra Space-dimension

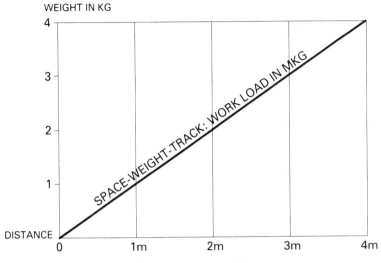

Fig. 4: Space-weight diagram

will reach B in one second, C in two, D in three, and E in four seconds. The seconds are marked vertically and the sound-distances horizontally. The diagonal indicates the position of each respective police post and the corresponding times and distances in which the sound will reach it. Based on Einstein/Hawking reasoning we can speak here of a **space-sound continuum (or even a space-time-sound continuum), elevating sound to the status of an extra space-dimension.**

The sun rises at point A at 4 a.m., when the thermometer indicates a temperature of 10° Celsius. At 6 a.m. the temperature will have risen to 15°, at 8 a.m. to 20°, at 10 a.m. to 25° and at midday to 30°C. The Celsius degrees are marked on the vertical line in units of height, corresponding with the mercury column of the thermometer. The times of day are marked horizontally. The diagonal line combines temperatures with the respective sun positions. Based on Einstein/Hawking reasoning we can speak

Has Hawking Erred?

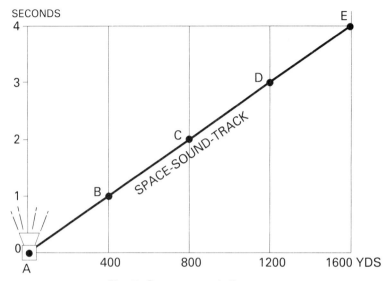

Fig. 5: Space-sound diagram

here of **a space-temperature continuum (or even of a space-time-temperature continuum), elevating temperature to the status of an extra space-dimension.**

What the comparative diagrams attempt to show is that, when closely examined, neither time, sound, temperature or weight, can be treated as extra space-dimensions. They must rather be considered as figures of speech only casually associated with space. Thus the expression 'in the space of five minutes', can be compared with such phrases as 'deep thought', 'being short of breath', or 'a broad smile'; none of them has anything to do with space measurement. Equally, 'a hot pursuit' has nothing to do with temperature and 'a heavy heart' has nothing to do with weight. Further, the phrases 'a soft heart' or a 'weak response' have nothing to do with physical hardness or weakness. A more extreme example is perhaps 'born with a silver spoon in his mouth', which has nothing to do

Time as an Extra Space-dimension

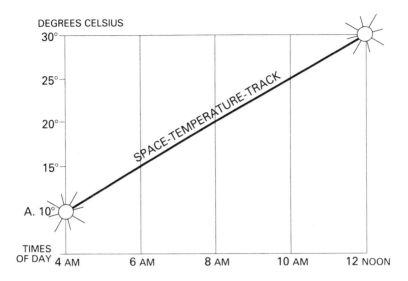

Fig. 6: Space-temperature diagram

with physical birth, but implies rather inherited wealth. Similarly, the phrases 'it took a long time for him to grasp the problem', or 'he waited only a short time', have nothing to do with metric distances. 'Long' and 'short' here strictly denote time duration recordable by clocks.

Comparing thus the actual diagrams in this manner, the phrases 'in the space of five minutes', or 'it took a long time' are not directly related to the measurement of a distance in linear space; nor does 'a rise in temperature' measured in degrees Celsius and marked in units of length along a thermometer imply any permanent space-temperature-relationship. The sound of an alarmbell alerting a distant police post does not imply a permanent sound-space-time relationship, nor that of weight to distance. Thus none of the above phenomena can play the part of an extra space-dimension, as all of them belong to different categories of natural phenomena which do not mix. It is the same principle which does not allow an intermingling of sound and temperature, temperature and weight and weight and sound.

12
The Laws of Transformation

Einstein, while trying to show that light behaves differently from conventional phenomena of nature (for example, sound), maintains that all measurements of time and distance in relation to light are variable. In the following examples, Einstein demonstrates his viewpoint (B42–45). I quote at length and provide the relevant illustrations.

Einstein's explanation (B41) bears reiteration:

> It is constantly necessary for the scientist when dealing with matter involving complex forms of motion (as in celestial mechanics, electrodynamics, etc.) to relate the magnitudes found in one system with those occurring in another. The mathematical laws which define these relationships are known as the laws of transformation.

Einstein explains (B42):

> The simplest transformation may be illustrated by a man promenading on the deck of a ship [see Example 1 below]: if he walks forward along the deck at the rate of three miles per hour and the ship moves through the sea at the rate of twelve miles per hour, then the man's velocity with respect to the sea is fifteen miles an hour; if he walks aft, his velocity relative to the sea is of course nine miles per hour.

Has Hawking Erred?

Example 1: Man Walking on Deck of Ship (1)

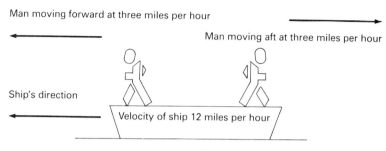

Static Sea Surface

Commenting on the above example Einstein writes (*Relativity*, 1988, p. 16): 'The law that we have just written down does not hold in reality'. And he adds: 'For the time being, however, we shall assume its correctness.'

Summation

1. Velocity of the **ship relative to the sea** (the static system) is twelve miles per hour.
2. Velocity of the **man relative to the ship** (the static system in this case) is three miles per hour.
3. Velocity of **man relative to the sea** (the other static system) when moving forward on ship is fifteen miles per hour, and when moving aft, nine miles per hour.

Example 2: Train Moving towards an Alarm Bell

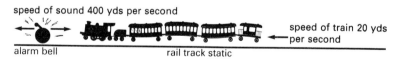

The speed of the sound in either direction is 400 yards per second.

← Direction of train, moving at a velocity of 20 yards per second.

The Laws of Transformation

Summation

1. Velocity of **train relative to track** (the static system), 20 yards per second.
2. Velocity of **sound relative to track** (the static system), 400 yards per second.
3. Velocity of **sound relative to train**, while moving towards the alarm clock is 420 yards per second, and after having passed the alarm clock, is 380 yards per second.

Einstein explains (B40):

> The sound waves produced by the bell spread away through the surrounding air at the rate of 400 yards a second. A railroad train speeds toward the crossing at the rate of 20 yards a second. Hence the velocity of the sound relative to the train is 420 yards a second so long as the train is approaching the alarm bell and 380 yards a second as soon as the train passes the bell. This simple addition of velocities rests on obvious commonsense, and has indeed been applied to problems of compound motion since the time of Galileo. Serious difficulties arise, however, when it is used in connection with light.

According to Barnett (B42):

> In his original paper on Relativity, Einstein emphasised these difficulties with another railway incident. Again there is a crossing, marked this time by a signal light which flashes its beam down the track at 186,284 miles a second – the constant velocity of light, denoted in physics by the symbol c. A train steams towards the signal light at a given velocity v. So by the addition of velocities one concludes that the velocity of the light beam relative to the train is c plus v when the train

moves toward the signal light, and c minus v as soon as the train passes the light.

This is projected in example 3 below.

Example 3: Train Moving towards Signal Light

Velocity c of light in either direction: 186,284 miles per second.
Velocity v of train moving towards signal light: 100 miles per second.

Summation

1. **Velocity of signal light c relative to track** (the stationary system) is 186,284 miles per second.
2. **Velocity v of train moving towards the signal light** is 100 miles per second relative to track (the stationary system).
3. **Velocity of light c, when measured relative to the velocity of the train** (which moves towards it), is 186,284 miles per second plus 100 miles per second, i.e. c plus v.
4. Velocity of light c measured relative to velocity of train (after having passed the signal light), is 186,284 miles per second minus 100 miles per second, i.e. c minus v. In both cases the rail track acts as a static system of reference.

Einstein maintains that this result conflicts with the findings of the Michelson-Morley experiment (carried out in 1887 by A. A. Michelson and E. Morley to detect the earth's motion through the postulated ether).
Barnett (B38) comments:

The Laws of Transformation

The one indisputable fact established by the Michelson-Morley experiment was that the velocity of light is unaffected by the motion of the earth. Einstein seized on this as a revelation of universal law. If the constancy of light is constant regardless of the earth's motion, he reasoned, it must be constant regardless of the motion of any sun, moon, star, meteor or other systems moving anywhere in the universe. From this he drew a broader generalisation, and asserted that the laws of nature are the same for all uniformly moving systems. This simple statement is the essence of Einstein's Special Theory of Relativity.

Returning to train example 3 above, Einstein points out that:

> Even if we imagine the train is racing towards the signal light at a speed of 10,000 miles per second, the principle of the constancy of the velocity of light tells us that an observer aboard the train will still clock the speed of the oncoming light beam at precisely 186,284 miles a second, no more, no less. The dilemma presented by this situation involves much more than a Sunday morning newspaper puzzle. On the contrary it poses a deep enigma of nature.

Einstein, says Barnett (B45)

> saw the problem lay in **the irreconcilable conflict** between his belief in (1) the constancy of the velocity of light, and (2) the principle of the addition of velocities, although the latter appears to **rest on the stern logic of mathematics, namely that two plus two makes four** [emphasis supplied].

Barnett writes further that Einstein

recognised in the former a fundamental law of nature. He concluded, therefore, that a new transformation rule must be found to enable the scientist to describe the relations between moving systems in such a way that the results satisfy the known facts of light.

An analysis of the argument fails however to disclose that any enigma is involved. Firstly, no one will dispute that in the preceding example **the actual velocity of light retains its old value of 186,284 miles per second, while its relative value vis-à-vis the speeding train is modified to 186,284 miles per second minus, or plus, 100 miles per second.** The problem is identical with the previous example of the train moving towards, or away from, an alarm bell. Here also the actual speed of sound remains unchanged, while its relative speed vis-à-vis the train is valued at 400 yards per second minus, or plus, 20 yards per second.

In order to point out the alleged flaws inherent in the old principles of the addition of velocities (which Einstein maintained to have discovered), Einstein advances another example (see B44):

> Once again he envisaged a straight length of track, this time with an observer C sitting on an embankment beside it. A thunderstorm breaks, and two bolts of lightning strike the track simultaneously at separate points, A and B. Now, asks Einstein, what do we mean by 'simultaneously'? To pin down this definition we assume that the observer is sitting precisely halfway between A and B, and that he is equipped with an arrangement of mirrors which enable him to see A and B at the same time without moving his eyes. Then if the lightning flashes are reflected in the observer's mirrors at precisely the same instant, the two flashes may be regarded as simultaneous.

The Laws of Transformation

Example 4: Einstein's Example of Simultaneity

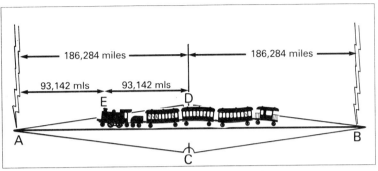

Train moving at velocity of light C

Total distance from A to B is two times 186,284 miles. Lightning bolt A will strike D's mirror when reaching point E halfway between A and C/D.

Einstein continues:

> A second observer is sitting precariously perched atop one of the cars with a mirror apparatus just like the one on the embankment. It happens that this moving observer finds himself directly opposite the observer on the embankment at the precise instant the lightning bolts hit A and B. The question is: will the lightning flashes appear simultaneous to him? The answer is: they will not. For if his train is moving away from lightning bolt B and towards lightning bolt A, then it is obvious that B will be reflected in his mirror a fraction of a second later than A. Lest there be any doubt about this, one may imagine temporarily that the train is moving at the impossible rate of 186,284 miles a second, the velocity of light. In that event, flash B will never be reflected in the mirrors because it can never overtake the train, just as the sound from a gun can never overtake a bullet travelling with supersonic speed. So the observer on the train will assert that only one lightning bolt struck the track. And whatever the speed of the train

may be the moving observer will always insist that the lightning flash ahead of him has struck the track first. Hence the lightning flashes which are simultaneous relative to the stationary observer are 'not' simultaneous relative to the observer on the train.

Barnett (B45) comments:

> The paradox of the lightning flashes thus dramatises one of the subtlest and most difficult concepts in Einstein's philosophy: the relativity of simultaneity. It shows that man cannot assume that his subjective sense of 'now' applies to all parts of the universe. For, Einstein points out, 'every reference body (or coordinate system) has its own particular time; unless we are told the reference body to which the statement of time refers, there is no meaning in a statement of the time of an event.' The fallacy in the old principle of the addition of velocities lies therefore in its tacit assumption that the duration of an event is independent of the state of motion of the system of reference.

Contrary to Einstein's interpretation (as explained by Barnett, B45), there seems to be no paradox involved in this example, nor does the illustrated relationship of 'simultaneity' dramatise one of the most difficult concepts in physics. Firstly, if scrutinised from a purely semantic viewpoint the terms 'now' and 'simultaneous' reveal themselves as belonging to a category of abstract words which are common in everyday speech. However, like other abstractions of the same order they lack specific meaning unless related to particular objects or situations. In the above case, they are alternatively linked to different 'systems of reference' – all this being in accord with our own, as well as Einstein's, views on relativity, previously elucidated.

The Laws of Transformation

To deal specifically with the abstract term 'simultaneous', it can be shown that just like the abstract 'space' it has multifarious meanings. One can speak of the mouth having the dual function of 'simultaneously' serving as a channel for the intake of food and as an inlet for breathing. The 'simultaneousness' experienced here reveals itself as a permanent condition. In chess one speaks of a 'simultaneous' event, when a chess master plays simultaneously against twenty opponents. After having opened with a first move against the first opponent, he then proceeds to the second one, again to make a first move. After having completed the first circuit against all the twenty, he will return to his first opponent for his second move, etc. In this case the 'simultaneity' involved extends over several hours, interrupted by the intervals during which the chess master moves from player to player. Another example is our solar system, insofar as the earth, while circling the sun, simultaneously revolves around its own axis. This is again a case where the simultaneousness is permanent.

In Einstein's train example (with the train moving at the speed of light) the stationary observer C who is located at the middle of the track will receive the lightning flashes from A and B reflected in his mirrors exactly at the same time. They will produce a situation which in accordance with Einstein's intended demonstration can be described as **truly 'simultaneous'**.

On the other hand, observer D, precariously perched on the roof of the train moving at the speed of light in the direction of A, will receive the mirror image of lightning flash A only when he or she has reached point E, which is pinpointed halfway between A and C. Furthermore, observer D will never receive the mirror image from flash B, since, as Einstein rightly points out, it can never reach D's mirror, because D, who is moving at the same speed (that of light) as the train, in advance of the flash, will already have reached point A at the end of the track,

Has Hawking Erred?

while the lightning flash B will only have arrived at the centre of the track at the point where observer C is located.

To a superficial observer it will appear almost as a certainty that both lightning flashes A and B reach the mirrors of both observers C and D 'simultaneously', at the very moment when they face each other at the centre of the track. However, close analysis reveals, as shown above, that this impression amounts to something like an optical illusion, and there is really no mystery involved. It simply happens that the illustrated case in example 4 involves only one 'simultaneity', namely the one relating to the stationary observer C, who is situated at the centre of the track. The other observer D, who is on the moving train, is in a different situation in respect to the lightning flashes; he perceives only one of them, namely A. Thus no simultaneity is involved at all.

Based on this supposedly paradoxical example of apparently comparative 'simultaneities', Einstein now pursues the argument further, concluding that the old principle of the addition of velocities has become obsolete. He says (and I repeat his previous comment):

> The fallacy of the old principle of the addition of velocities lies in its tacit assumption that the duration of an event is independent of the state of motion of the system of reference.

In further pursuit of the argument it will be shown that Einstein's above assumptions do not match the facts of the situation. In questioning them, I submit some more illustrated examples.

Example 5: Two trains moving at different velocities on parallel tracks:

Train A/1, velocity 186,000 miles per second

rail track stationary

The Laws of Transformation

Train A, which is moving at the velocity of 40 metres per minute, is overtaken by train B moving parallel to it at a speed of 50 metres per minute. Now, while train B is overtaking train A, a timekeeper with a stopwatch is perched on the roof of train A, registering the velocity of train B over a certain measured distance, while he is being overtaken by it. The result is that the velocity of train B relative to train A will measure exactly ten metres per minute, resulting from the subtraction of train A's velocity of 40 metres per minute from train B's velocity of 50 metres per minute.

Example 6: Two trains moving on parallel tracks:

Train B/1, 186,284 miles per second

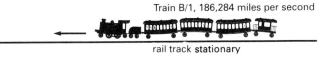

rail track stationary

Summation

1. **Actual velocity of A relative to stationary track** is 40 metres per minute.
2. **Actual velocity of B relative to stationary track** is 50 metres per minute.
3. **Relative velocity of train B in respect to train A** while overtaking it amounts to 10 metres per minute (i.e. 50 minus 40 metres per minute).

What needs stressing in this example is that the measurements of the above three values are related in each case, in respect of the timing of the duration of each event to regular conventional calendar time; and in respect of measuring distance to our conventional metre. Both sets of circumstances are again related to the fixed system of reference presented by the stationary railway tracks.

Has Hawking Erred?

To show that moving systems advancing at the velocity of light are subject to the same criteria of relativity as systems moving at a lower velocity, the following example 6 serves as a pointer.

One train, B/1 is moving at the velocity of light, the other, A/1, at the lesser velocity of 186,000 miles per second.

While train A/1 is moving at the velocity of 186,000 miles per second (which is 284 miles per second less than the velocity of light), it is overtaken by train B/1, which is moving at a velocity of 186,284 miles per second, equalling the velocity of light. Apart from the difference in velocities, example 6 is in all respects identical with example 5.

Summation

1. **The actual velocity of train A/1 relative to its stationary track** is 186,000 miles per second. (284 miles per second lower than the velocity of light.)
2. **The actual velocity of train B/1 relative to its stationary track** is 186,284 miles per second.
3. **Velocity of train B/1 relative to train A/1** (while overtaking the latter), is 284 miles per second.

As in the preceding example 5, the measurements of all the three values are based, in respect of their duration, on regular conventional calendar time, and in respect of the measurement of distance, on our conventional metre. Both time and distance measurements are again related to the fixed system of reference presented by the stationary railway tracks.

The important conclusion to be drawn from examples 5 and 6 is that in retrospect, all the preceding examples dealing with the 'addition of velocities' are subject to the same relationships of time and distance measurements as

The Laws of Transformation

detailed in the preceding paragraph. Furthermore, contrary to Einstein's assumption, in none of his, or our quoted examples, can it be shown that the measurements of time, based on conventional calendar time (i.e. the duration of events), or the measurements of space, i.e. of linear distance, based on the conventional metre, are variable quantities.

To show once more that the velocity of light cannot claim any exemption from this rule, I quote two more examples.

Example 7: Two Trains, C and D, Travelling in Opposing Directions on Parallel Tracks at the Speed of Light

Summation

1. **The actual velocity of train C relative to the stationary track** is 186,284 miles per second.
2. **The actual velocity of train D (moving in the opposite direction) relative to the stationary track** is also 186,284 miles per second.
3. **The relative velocity of train C to train D, measured from on board train** D, while passing the former, is 186,284 miles per second plus 186,284 miles per second, resulting in the added velocities of C and D, totalling 372,568 miles per second. This is double the normal velocity of light. In the reverse assessment of the relative velocity of train D, measured on board train C while passing train D, the same result is obtained.

Einstein has generally maintained that the velocity of light can not be exceeded even if it is assessed relative to

Has Hawking Erred?

other moving systems. However, at some later stage of his enquiry he seems to have recognised that 'there are certain types of motion that relative to the observer will have speeds faster than light'. (Milton Rothman: 'Things That Go Faster Than Light', *Scientific American*, July 1960.)

To question further Einstein's proposition that the speed of light relative to another system can never exceed 186,284 miles per second, I advance the simple example of an ordinary light bulb.

Example 8: Light Bulb

A light bulb LB, in a static central position, is sending light waves R and L to the right and to the left.

Summation

1. **The actual velocity of light waves R relative to LB (the static system)** will be 186,284 miles per second.
2. **The actual velocity of light waves L relative to LB (the static system)** will equally be 186,284 miles per second.
3. **The relative velocity of light waves L measured against the actual velocity of light waves R** will amount to the added velocities of L and R, namely to 372,568 miles per second, which is twice the normal velocity of light. No proposition based on the Michelson-Morley experiment can alter this fact.

To present more facts intended to emphasise Einstein's viewpoint, I refer to his re-examination of the boat example previously illustrated under example 1, (see

The Laws of Transformation

p. 84), but now provided by Einstein with the additional details of clocks.

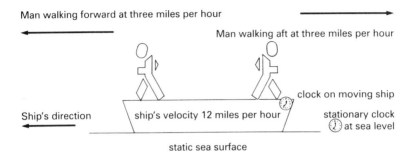

Example 9: Man Walking on Deck of Ship (2)

Einstein elucidates:

> In the case of the man pacing the deck of a ship (1), it was assumed that if he walked three miles in one hour as timed by a static clock on the moving ship his rate of progress would be just the same as that timed by a stationary clock anchored somehow in the sea. It was further assumed that the distance he traversed in one hour would have the same value whether it was measured relative to the deck of the ship (the moving system) or relative to the sea (the stationary system). This constitutes the second fallacy in the addition of velocities – for distance, like time, is a relative concept, and there is no such thing as a **space interval** independent of the state of the motion of the system of reference [emphasis supplied].

In a special reference to the 'Lorentz transformation', Einstein (*Relativity*, 1988, p. 41), drew attention to the fact:

> that a theory of this phenomenon was given by H. A. Lorentz [Dutch physicist, 1853–1928] long before the

statement of the theory of relativity. This theory was of a purely electrodynamical nature, and was obtained by the use of a particular hypothesis as to the electromagnetic structure of matter. This circumstance, however, does not in the least diminish the conclusiveness of the experiment as a crucial test in favour of the theory of relativity, for the electrodynamics of Maxwell-Lorentz [J. C. Maxwell, Scottish physicist, 1831–79], on which the original theory was based, in no way opposes the theory of relativity. Rather has the latter been developed from electrodynamics as an astoundingly simple combination and generalisation of the hypotheses, formerly independent of each other, on which electrodynamics was built.

The following abstract of the Lorentz transformation will enable the knowledgeable reader to make his own assessment of the thesis.

13
The Lorentz Transformation

Einstein (B46–47) gives this description:

The Lorentz transformation relates distances and times observed on moving systems with those observed on systems relatively at rest. Suppose, for example, that a system, or reference body, is moving in a certain direction, then **'according to the old principle of velocities'** [emphasis supplied], a distance or length x, measured with respect to the moving system along the direction of motion, is related to length x, measured with respect to a relatively stationary system, by the equation $x' = x + vt$, where v is the velocity of the moving system and t is the time. Dimensions y' and z', measured with respect to the moving system at right angles to x' and at right angles to each other (i.e., height and breadth), are related to dimensions y and z on the relatively stationary system by $y' = y$ and $z' = z$. And finally a time interval t, clocked with respect to the moving system, is related to time interval t', clocked with respect to the relatively stationary system, by $t' = t$. In other words, distances and times are not affected, in classical physics, by the velocity of the system in question. But it is this presupposition which leads to the paradox of the lightning flashes. The Lorentz transformation reduces the distances and times observed on moving systems to the conditions of the stationary observer, keeping the velocity of light c a constant for all observers. Here are the equations of the Lorentz transformation which have supplanted the older and evidently inadequate relationships cited above:

$$x' = \frac{x - vt}{\sqrt{1 - (v^2/c^2)}}$$
$$y' = y$$
$$z' = z$$
$$t' = \frac{t - (v/c^2)x}{\sqrt{1 - (v^2/c^2)}}$$

It will be noted that, as in the old transformation law, dimensions y' and z' are unaffected by motion. It will also be seen that if the velocity of the moving system v is small relative to the velocity of light c, than the equations of the Lorentz transformation reduce themselves to the relations of the old principle of the addition of velocities. But as the magnitude of v approaches that of c, then the values of x' and t' are radically changed.

An examination of Einstein's account (**preceding the abstract of the Lorentz transformation**) shows not only that it is a complex and enigmatic discourse, but that it is at times quite dubious. For example, the phrase '**space interval**' (on p. 97) can only mean that the interval is a time measure, depicting the duration of an event, because the normal measurement of linear space distance is as a rule made in metric units of length, width or height, which are independent of time and have no duration in terms of time. Only time has duration, and its measurement is always based on units of conventional calendar time, measured by clocks on earth. Einstein's use of the

The Lorentz Transformation

phrase '**space interval**' in the present context must therefore be considered highly ambiguous. A space interval correctly interpreted would normally describe a gap in a linear distance.

When it comes to the measurement of the man's rate of progress measured by a clock fixed on the moving boat and another clock anchored at sea, with both supposedly showing the same duration of time, it can only mean that this time measurement refers to the total time having elapsed during which the man travelled on the boat, covering the entire journey. Its value is not given in Einstein's original example. However, this makes no difference to the result, because the duration of the man's time spent on the ship is not directly related to the additional distance he covered while walking on the ship.

An equivalent example would be that of a man travelling on a train from London to Liverpool. The man's wristwatch would register the same time interval between departure and arrival as would the station clocks at London and Liverpool. Furthermore the total distance the man travelled from London to Liverpool would remain the same whether the man walked some distance within the train or whether he remained seated during the entire journey.

Basically, therefore, the first part of Einstein's account demonstrates nothing particularly new or extraordinary. Neither does it dislodge the old principle of the addition of velocities, the correctness of which will be further demonstrated in example 10. In the progress of the argument it will also be shown that for all practical purposes of physical measurement, **time and distance do not represent variable quantities, irrespective of how Einstein interprets the Lorentz transformations**. In all the examples given, measurements are related to the fixed values of conventional calendar time and the conventional standards of the metric system applied retroactively over

Has Hawking Erred?

billions of years. They show no variation in their regularity, whether they are used to measure static distances or objects in motion.

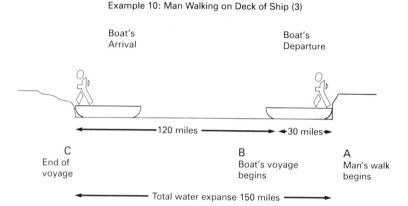

Example 10: Man Walking on Deck of Ship (3)

I must explain here, that in order to convey a clear picture of Einstein's argument, I have found it necessary to exaggerate the distances in (3) and use an imaginary boat measuring 30 miles in length. What the illustration proves is that the distance the man walks in 10 hours on the deck of the boat is 30 miles (at 3 miles per hour) in covering the length of the boat from A to B. During the same interval of 10 hours, the boat will cover a distance of 120 miles from B to C (at a velocity of 12 miles per hour): both man and boat will arrive at the same time at point C. Moreover, when measured relative to the stationary sea (not measured by a clock but by a metric measure), the man will, during the same ten hours, have covered a total distance of 150 miles (stretching from A to C). Yet when measured relative to the boat (in metric terms), the man covers a distance of 30 miles during the same ten hours. This example negates Einstein's preceding assertion that **the relative distances** (in (2)) travelled by boat and man are identical. On the other hand, **the actual**

The Lorentz Transformation

distances travelled by boat and man are of course identical.

Einstein asserts, and I quote again:

The scientist who wishes to describe the phenomena of nature in terms that are consistent for all systems throughout the universe **must regard measurements of time and distance as variable quantities**. The equations comprising the Lorentz transformation do just that. They preserve the velocity of light as a universal constant, but modify all measurements of time and distance according to the velocity of each system of reference [emphasis supplied].

This particular phrasing seems to disregard the effect which the absolute velocity of light exercises on the measurements of time and space, since both are just as much absolute unvariable entities as is the velocity of light itself, simply because the velocity of light is measured in the same units (i.e. of conventional seconds). **The constant measure of a light second and a second of 'regular calendar time' are the same; while the distance in a light-related metre (which represents a definite fraction of a second during which the light travels to cover one metre's distance), is identical to, and synchronises with, our earth metre. All three, 'speed of light', 'regular calendar time' and the 'earthly metre' are strictly interrelated.**

If the velocity of light is to be preserved as a universal constant, then the regular second of calendar time and the regular distance in metres at which the velocity of light is expressed (both of them being basic ingredients of the constancy of light) must also be universal constants. Otherwise, the assertion of the constancy of light breaks down.

The manner in which Einstein formulates the above

quotation not only conceals this breakdown, but makes his conclusions questionable.

Hawking (p. 168), in referring to the original impact of Einstein's theory of relativity on the physics establishment, writes that Eddington [British astrophysicist, 1882–1944] is supposed to have proclaimed seventy years ago that only two people could make sense of Einstein's general theory of relativity. Nowadays, says Hawking, thousands of university students can make sense of it, while millions of people have at least an idea what it is all about.

I dare to express my doubts about this statement, because although some students of relativity may have pretended to have grasped all of it, they certainly can only have understood part. Some of Einstein's phraseology is often involved and at times quite enigmatic so that even today few physicists can make sense of everything Einstein pronounced. This does not however, discredit the main corpus of his monumental achievements.

14
Space-time in Voluminal Space

Universal voluminal space is a natural phenomenon, non-material and of limitless dimensions, within which all material objects coexist. The space distances between these objects, whether they are in motion or at rest, are measurable in linear metric units. All material objects again are three-dimensional and possess volume. In dictionary terms volume is defined as the amount of space (often expressed in cubic units) that a three-dimensional object occupies; in other words, three-dimensionality stands for volume and volume for three-dimensionality.

This three-dimensional quality of all objects in the world is axiomatic, insofar as there cannot exist any material thing with either more or less than three dimensions. And this three-dimensionality is just as absolute as the three-sidedness of a triangle. And just as a triangle can never break out of its three-sided frame, so can material objects never escape from their three-dimensional prison, nor squeeze in additional dimensions.

Einstein and Hawking, nevertheless, in common with other physicists, have postulated the existence of the new concept of space-time, elevating time to the status of an extra space-dimension. They claim that time can play the part of a fourth dimension within what was formerly

Has Hawking Erred?

restricted to an uncompromising three-dimensional space configuration.

Now, while it is impossible to prove mathematically, even on paper, that a triangle can have more than three sides, the theory of relativity claims to have proved mathematically (though solely on paper), the existence of a four-dimensional space-time continuum, implying thereby that material objects, normally confined to three-dimensionality, can have more than three space-dimensions. This postulate violates one of the most fundamental laws of nature, thus showing that there are occasions when mathematics can lose its compatibility with physical reality. (For more details on this see preceding chapter 3 Mathematical Vulnerability).

Yet this possibility was inadvertently admitted by Einstein when he wrote (B59):

> Since time is an impalpable quantity, it is not possible to draw a picture or construct a model of a four-dimensional space time continuum. But it can be imagined and it can be presented mathematically.

Hawking (P24) writes: 'It is impossible to imagine a four dimensional space called space-time. I personally find it hard to visualize three-dimensional space.' Yet surprisingly, and against his own better judgement, he presents us with several four-dimensional space-time diagrams (pp. 26 and 27 of his book). The result is doubtful, because all of them appear incomprehensible; although most readers, struck with awe at such profound wisdom, may uncritically accept them at their face value, or else will pretend comprehension.

One need not, however, be deeply versed in modern physics or mathematics to form a sane judgement about an extra, fourth space-dimension, because in physical reality, accepted without prejudice, it is absolutely axio-

Space-time in Voluminal Space

matic that voluminal space or space objects ('voluminal' simply being another term for 'three-dimensional') can have only three dimensions.

While Euclid could certainly not have intended to postulate a thing called space-time, he demonstrated more than two millennia ago, though solely in writing, that one can generate space of four dimensions mathematically; furthermore, there is no need to stop at four. In extended Euclidean geometry we can go on to space of five, six, seven or more dimensions. In developing the general theory of relativity Einstein found it necessary to adopt a four-dimensional general Riemannian geometry. Instead of a fourth, strictly geometrical space-dimension, Einstein made time his fourth dimension. Einstein pointed out that there is nothing mysterious in this concept. It merely means that every event that takes place in the universe is an event occurring in a four-dimensional world of space-time.

The most telling example of a mix-up between linear and voluminal space – which resulted in the concept of space-time as a universal constant – leads us back to Minkowski. It is well known that Einstein derived the main impetus of time, functioning as an extra space-dimension and resulting in space-time, from his mathematics teacher, the Russian-German geometrician Hermann Minkowski (1864–1909). Minkowski had been one of Einstein's teachers at Zurich Polytechnic. His fundamental thesis was that space and time had to be considered together as a single entity, forming a four-dimensional space-time complex. In 1908 he announced in a famous lecture at the University of Göttingen that 'Henceforth space by itself, and time by itself, are doomed to fade away into mere shadows, and only a kind of union of the two will preserve independent reality.' (It is not known whether Minkowski's 'space' referred specifically to 'linear', two-dimensional space, to three-dimensional 'voluminal space' or to 'space in general'.)

Einstein refers thus to Minkowski's discovery (*Relativity*, pp. 56–7): '**Pure space distance** of two events with respect to K' (K' being a first coordinate-system) **results in "time distance"** of the same event with respect to K' (K' being a second coordinate-system)' [emphasis supplied].

Although Minkowski and Einstein based their propositions originally on linear space projections (as illustrated by Einstein's example of the New York–Chicago express train) both Minkowski and Einstein subsequently transferred this linear space-time concept into the realm of three-dimensional voluminal space. This purely speculative transfer (an obviously philosophical exercise) from one space medium into another (i.e. linear into voluminal) was remarkable insofar as it subsequently involved the adoption of any number of additional space-dimensions, the transfer of linear space-time into voluminal space and the curving of voluminal space and the treating of universal voluminal space as if it were a plastic medium. This also included the identification of space with time and in reverse that of time with space, and so forth. For example (B60), Einstein speaks of 'the equivalence of space and time'. Introducing his proposition, Einstein writes (E55):

> The non-mathematician is seized by a mysterious shuddering when he hears of 'four-dimensional' things, by a feeling not unlike that awakened by thoughts of the occult. And yet there is no more commonplace statement than that the world in which we live in is a four-dimensional space-time continuum.

Commenting on this wide-ranging assertion, Barnett (B57) consentingly adds:

> Once the meaning of the word 'continuum' is properly

Space-time in Voluminal Space

grasped, Einstein's picture of the universe as a four-dimensional space-time continuum – and this is the view that underlies all modern conceptions of the universe – becomes perfectly clear. **A continuum is something that is continuous** [emphasis supplied].

Einstein adds further (E52):

The inherited prejudices about time and space did not allow any doubt to arise as to the prime importance of the Galilean transformation for changing over from one body of reference to another. Now assuming that the Maxwell-Lorentz equations hold for a reference body K, we then find that they do not hold for the reference body K' moving uniformly with respect to K, if we assume that the relations of the Galilean transformation exist between the co-ordinates of K and K' . . . [E56] As a consequence pure **'space-distance'** of two events with respect to K, **results in time-distance** of the same events with respect to K'. But Minkowski's discovery, which was of importance for the formal development of the theory of relativity, does not lie here. It is to be found rather in the fact of his recognition that the four-dimensional space-time continuum of the theory of relativity, in its most essential formal properties, shows a pronounced relationship to the three-dimensional continuum of Euclidean geometrical space [emphasis supplied].

What we immediately encounter here is a dilemma between reality and theory (i.e. the theory of relativity), because it can be established, as previously shown, without reasonable doubt that time and space, are two entirely different phenomena (the one an artificial human creation and the other an intrinsic fact of nature), cannot intermingle. The one, 'space-distance', is strictly limited to linear space distances between objects measurable in metric units of length; while the other, 'time-distance', can only express duration, measurable in seconds, minutes and hours on the face of a clock. An interchange or fusion of the two concepts is incompatible with natural events.

Even in Einstein's own example, given below, we find that the space co-ordinates he mentions can only refer to linear space-distance, the measurement of which is independent of time. Having once established the position of a body with the help of these co-ordinates, however, a time relationship can then be established with another body in motion, the position of which is also determined by linear co-ordinates.

To explain the system of co-ordinates (E 5–8) Einstein writes:

> Every description of the scene of an event or of the position of an object in space is based on the specification of the point on a rigid body (body of reference) with which that event or object coincides. This applies not only to scientific description, but also to everyday life.

His example is Trafalgar Square in London:

> If I analyse the place specification 'Trafalgar Square, London', I arrive at the following result. The earth is the rigid body to which the specification of place refers; 'Trafalgar Square, London', is a well-defined point, to which a name has been assigned, and with which the event coincides in space.

There then follows a cloud example:

> If, for instance, a cloud is hovering over Trafalgar

Square, we can determine its position relative to the surface of the earth, erecting a pole perpendicularly on the Square, so that it reaches the cloud. The length of the pole measured with the standard measuring-rod, combined with the specification of the position of the foot of the pole, supplies us with a complete place specification. On the basis of this illustration, we are able to see the manner in which a refinement of the conception of position has been developed.

(a) We imagine the rigid body, to which the place specification is referred, supplemented in such a manner that the object whose position we require is reached by the completed rigid body.

(b) In locating the position of the object, we make use of a number (here the length of the pole measured with the measuring-rod) instead of designated points of reference.

(c) We speak of the height of the cloud even when the pole which reaches the cloud has not been erected. By means of optical observations of the cloud from different positions on the ground, and taking into account the properties of the propagation of light, we determine the length of the pole we should have required in order to reach the cloud.

Einstein continues:

From this consideration we see that it will be advantageous if, in the description of position, it should be possible by means of numerical measures to make ourselves independent of the existence of marked positions (possessing names) on the rigid body of reference. In the physics of measurement this is attained by the application of the Cartesian system of co-ordinates.

This consists of three plane surfaces perpendicular to each other and rigidly attached to a rigid body. Referred

Has Hawking Erred?

to a system of co-ordinates, the scene of any event will be determined (for the main part) by the specification of the lengths of the three perpendiculars or co-ordinates (x, y, z) which can be determined by a series of manipulations with rigid measuring-rods performed according to the rules and methods laid down by Euclidean geometry.

In practice, the rigid surfaces which constitute the system of co-ordinates are generally not available; furthermore, the magnitudes of the co-ordinates are not actually determined by constructions with rigid rods, but by indirect means. If the results of physics and astronomy are to maintain their clearness, the physical meaning of specifications of position must always be sought in accordance with the above considerations.

We thus obtain the following result: every description of events in space involves the use of a rigid body to which such events have to be referred. The resulting relationship takes for granted that the laws of Euclidean geometry hold for 'distances', the 'distance' being represented physically by means of the convention of two marks on a rigid body.

What must immediately come to the reader's mind is firstly, that Trafalgar Square, London, cannot count as a well defined 'point', but is a large surface which is at places of irregular height and width. Secondly, to use a floating cloud (because no cloud can be described as static) as a specific event above Trafalgar Square which can be located (symbolically) by an exactly measured rigid pole, is a rather questionable decision. A cloud of any description is an irregular vaporous entity that is bound to change its shape or volume practically from moment to moment. To measure its height (or rather that of its underside) seems an altogether precarious venture. One has to conclude therefore that the cloud example is ill-chosen.

Space-time in Voluminal Space

However, despite the questionable choice of Einstein's space-co-ordinate example, certain facts are apparent: all measurements of his space-co-ordinates are related to linear space-distances; and no time relationships are attached to any of the co-ordinates, in spite of the fact that Einstein categorically maintains that all measurements in space depend on measurements of time (see B60). It is only when the relative position of different systems in motion is considered that the time factor becomes involved.

Generally, Einstein's examples given above have to be viewed with the following facts in mind:
1. All material objects in the universe are three dimensional, including the objects whose position is specified in Einstein's co-ordinate systems of K and K'.
2. To determine the position of any material object in voluminal space, linear space measurement in the form of metric units of length must be employed. This is of course also obligatory when measuring the distances of each of the three space co-ordinates.
3. None of the linear distance measurements so far cited shows the need for any time relationship, nor does the process of measuring and determining the position of the object within voluminal space (including K and K'), influence the dimensional qualities of the objects measured, insofar as their three-dimensionality remains unimpaired. Neither is the empty space (although it may be filled with air, ether or a gravitational field), within which the measurement is carried out, affected in any way by the process of the measurement itself. The example of a football kicked into the space above a football field may serve as an illustration. The football, being a three-dimensional space object, will not only not change its three-dimensional nature while flying through the space above, but will also leave no impact on the voluminal space

Has Hawking Erred?

through which it travels – the latter playing a totally passive role in the process.

4. So far the time factor has not been mentioned. Let us now assume that we want to determine the position of the football when reaching its highest point above the football ground while travelling through the air. Let us assume that this happens at a special point (i.e. Einstein's point of event), exactly ten metres above the ground at one hundredth of a second after four o'clock in the afternoon on a specific date. What now transpires is as follows:

The timing of the event, as can be observed, makes absolutely no impact on the three linear space co-ordinates that served to measure the event and determine its position in space; neither does the timing of the event have any impact on the flight path of the football itself, nor on its three-dimensionality. Furthermore, if we measure the linear distance of the flight path which the football describes while being off the ground, we cannot discover any change in the one-dimensionality of measurement, which is strictly linear. Only if we measure the duration of the time which the football spends in the air before reaching the ground does the time factor come into play. The question then is, how does the time factor influence any particular aspect of the entire transaction from the moment the football is kicked into the air up to the time when it lands back on the ground (including the determination of its highest point of flight)? The answer is that the time factor exercises no influence whatsoever on the football's spatial aspects except in recording the duration and possibly the speed of its flight. This recording of the duration of the flight on a clock is an entirely passive act, following the same principle which generally applies to the measurements of linear distances or of temperatures. Neither the metre-rod used in the distance measurement of the flight path of the body, nor the thermometer used

Space-time in Voluminal Space

in the temperature measurement of an object, leaves any impact on the objects measured. As to the time recording the duration of the motion of an object as it covers a linear distance, this exercises no influence on the space-dimensionality of the object timed, and more specifically does not change its voluminal status. Time measurement is simply a process of recording time duration measured by a clock. It has no other function.

In the face of such simple truths the postulation that time, or in other words, the recording of time duration, can function as an extra space-dimension, thereby becoming part of the new concept of space-time and in the context of space-curvature can even bend as if it were a material substance, does not conform to physical reality and ought therefore to be rejected. No mathematical manipulation, however elaborate, can alter this fact.

I conclude by suggesting that the above football example can be used comparatively to determine the position of other events in space and in time, whether they happen on earth or in any other region of the universe. In these examples also, the concept of space-time continues to be inapplicable.

15
The Enigma of the Curvature of Space

It is difficult to find in the writings of Einstein and Hawking a rational explanation of what they exactly mean when they speak of 'the curvature of space'. We find no clear clue to its how and why. When it emerges, its sudden entry is like that of a stranger from outer space.

Already the plain phrase 'curvature of space' (treating space like a material object), is quite enigmatic, because although both authors have disposed of ether, which would give voluminal space at least some semblance of reality (equalling ether with empty space), they somehow associate space-curvature with gravity. When further linked with the time factor, as when they speak of the curvature of time and space, the phrase becomes even more mysterious; although in the explanation of space-curvature that follows the time factor seems to play no essential part.

The two scientists surprise us by suddenly bursting on the scene with such statements as (Einstein, B29):

> Gravity is not a force like other forces, but a consequence of the fact that **space-time is not flat**, as has previously been assumed, **but is 'curved' or 'warped'** by the distribution of mass and energy in it [emphasis supplied]

The Enigma of the Curvature of Space

while Hawking (p. 33) points out that:

before 1915, it was natural to think that space and time went on forever. Thus bodies moved, forces attracted and repelled, but time and space went on forever. The situation is now different. In the general theory of relativity space and time are now dynamic quantities; when a body moves, or a force acts, **it affects the curvature of space and time** [emphasis supplied].

Neither man tells us however, how 'the curvature of space' is achieved nor what the phrase actually means. Do they mean the curvature of linear space, a principle which shows at least some linkage to reality, or the curvature of voluminal space, which appears to be totally irrational?

To trace the origin of such notions, one has to refer back to Minkowski, who can be considered the originator of the space-time principle. In this respect I have to reiterate Minkowski's dictum that space and time have to be considered as a single entity – a four-dimensional space-time – leading to the categorical statement that only a union of the two, that of space and time, will preserve an independent reality.

This, in my opinion, purely speculative and rather philosophical assertion, unaccompanied as it is by any valid explanation of how the fusion of space and time is actually accomplished, and of what Minkowski exactly means when he speaks of 'space' and 'time' (insofar as he fails to define these terms), was uncritically accepted by Einstein and subsequently by the physics establishment, including of course Hawking. It became a categorical imperative, leading to the further assertion that space and time could bend, warp, curve and even flatten as if they were material substances.

To point to the absurdity of such assumptions I will

Has Hawking Erred?

recall some facts about space and time which are relevant here.

Firstly, linear one-dimensional space and two-dimensional surface space can only gain reality when applied to the measuring of actual objects. Without such an association they are merely descriptive phrases without any physical presence, expressing solely attributive properties. Thus, unless a metric measure is used to measure the distance (whether straight or curved) between objects, 'one-dimensional space' must remain an abstract phrase. And the same holds for two-dimensional space.

Pure abstract 'space' in any of its three forms – one-dimensional linear, two-dimensional surface and even three-dimensional voluminal – are descriptive expressions which denote the measurement of linear distance, surface and volume, all three of them being merely abstractions. They cannot be treated like material objects and therefore cannot be subject to any physical manipulation. They cannot bend, warp or flatten, nor can they register temperature, weight or gravitational effects. Only material space objects, whether solid, fluid or gaseous, can be subject to physical influences. On the other hand all material objects in the universe are three-dimensional space objects and can therefore be subjected to bending or other forms of spatial distortion.

Let us now reconsider the abstract entity of universal voluminal space. It houses all the objects that exist in the universe, from the biggest galaxies down to the tiniest sub-atomic particles of matter, all of these being three-dimensional voluminal space objects. What role does one-dimensional linear space, whether straight or curved, play in this arrangement? The answer is (and I repeat): it is only a descriptive phrase which denotes a distance between material objects, or if objects are in motion, it denotes the distance they travel. However, such measurements only gain validity when related to a fixed frame of reference.

The Enigma of the Curvature of Space

Thus, the abstraction 'linear space' only gains meaning when used in association with material objects. Without such an association it is nothing and remains a purely semantic word construction.

Furthermore, all distance measurements occur within the bounds of universal voluminal space, which, as we have shown, is the hypothetical void which houses the universe. In this scenario, universal voluminal space plays the part of a passive medium, just as the voluminal space over a football field, through which a football is repeatedly kicked, remains unaffected by the play going on within its bounds.

The abstract term 'space' only gains reality in reference to material things. And as all material objects which exist in the universe, whether solid, fluid or gaseous, are three-dimensional voluminal space objects, they can all (in contrast to the former abstractions) be subjected to spatial distortions like bending, warping, flattening, shrinkage and expansion, and to the effects of temperature, weight or gravity. Unless it can therefore be shown that the term 'space' standing on its own can have a material existence – and the three purely abstractive space concepts, one-dimensional, two-dimensional and three-dimensional, do not have – the assertion that they can be subject to spatial distortion, in their abstract state is, and must remain, unanswerable.

Furthermore, in their abstract, non-material form, none of the three spatial concepts have any specific time relationship – just as a line drawn on a piece of paper (presenting a one-dimensional linear space-example) is indifferent to time. Neither has the voluminal space above a football field any specific time relationship, irrespective of whether or not a game of football is going on within its bounds or not.

Has Hawking Erred?

The role of time vis-à-vis space

In chapter 5, I showed that the purely abstract term 'time' lacks any meaning unless associated with the measurement of the time duration between physical events or the age determination of material objects. However, such time measurements only gain practical use, both in everyday life and in science, if related to our conventional terrestrial calendar, which again is strictly attuned to the regular astronomical movement of the earth around its own axis and to its encirclement of the sun.

It is remarkable that, although our terrestrial calendar is only a few thousand years old, it is retroactively applicable to events which date back to the assumed creation of the universe. This has been estimated to have occurred between ten and twenty billion years ago, an estimate that has now possibly been pushed further back to a hundred billion years or more.

There is even now a theory that postulates that the cosmos is trillions of years old. An article in *Time* magazine (2 September 1991), entitled 'The Big Bang Under Fire', suggests that 'a few scientists believe it is the Big Bang theory itself that is wrong. Instead of a universe that exploded into being 20 billion years ago and grew by way of gravity's tug, they postulate a cosmos trillions of years old and shaped not by gravity but by electricity and magnetism'.

Not only do these billions and even trillions of years represent a regular flow of calendar years all of equal duration, but they retroactively embody time divisions of the tiny fractions of seconds. Here I recall the Big Bang (the creation of the universe) theory, which proclaims (P424–5) that one thousandth of a second after creation there emerged a primordial fireball which can be described by a well-established physical theory. What is important to mention here is that these primordial parts of a second

The Enigma of the Curvature of Space

are identical with the fractions of our own conventional second.

Even more remarkable in this reference is that the regular flow of calendar years (or even that of split seconds), being an arbitrary, man-made, comparative measuring device, goes on relentlessly and pays no heed whatsoever to events, which simultaneously happen throughout universal voluminal space. Equally, the earth/sun cycle on which our time standard is based, is paralleled by earthbound clocks, which again, while recording the regular time flow, pay no heed to events happening in the universe. In this context, any time relationship to events is only casual, serving specific human needs for solving practical problems.

What must now be obvious to all is that time in the above context, being a purely abstract, comparative medium which denotes a regular time flow, has no material existence of its own and can therefore not be subject to any physical manipulation. Yet Einstein and Hawking, while considering time as an ingredient of space-time, allegedly subject to bending and other physical distortions, assert that time can bend and that space, being equivalent to time, must therefore also be subject to slowing down. Thus, in space-time, both time and space are treated as if they were physical substances. This further assumes indirectly that the earth/sun relationship is subject to bending, and that on a smaller scale the space in my room can slow down, or in a contrary situation can speed up. That such theories do not match reality is incontestable. The notion that time can bend and voluminal space can slow down can only be described as fantasy.

Now, while the slowing of space and the bending of time defy physical reality, I am quite sure that there are skilful mathematicians who by means of some complicated equations are able to prove that the above actions

Has Hawking Erred?

are theoretically possible. As a matter of fact such equations are part and parcel of the space-time theory.

My endeavour to learn how the concept of the curvature of space originated is outlined below, though I admit I may have missed the gist of the problem. Firstly, to repeat the homily by Einstein (B57) cited on page 110:

> A non-mathematician is seized by a mysterious shuddering when he hears of 'four-dimensional' things, by a feeling not unlike that awakened by thoughts of the occult. And yet there is no more commonplace statement than that the world in which we live is a four-dimensional space-time continuum . . . Once the meaning of the word 'continuum' is properly grasped the picture of the universe as a four-dimensional space continuum – and this is the view which underlies all modern conceptions of the universe – becomes perfectly clear.

According to Einstein, a ruler or a railway track are examples of one-dimensional space continuums. But as we have pointed out before, both are static entities which are independent of time. And although Einstein defines a flat surface as a two-dimensional continuum and voluminal space as a three-dimensional one, no true continuity is demonstrable in either, nor can one discover any link with what is called the 'curvature of space'. The problem becomes even more protracted when we move from a one-dimensional static situation to a volatile one, involving the movement of objects.

It is quite obvious that a train and a railway track (both being separate entities) progress in a continuous line. Actually, the train line is only hypothetically continuous, because in real life the train leaves no continuous trace in the space through which it travels. Equally so, a ship moving on the surface of the sea, although also travelling

The Enigma of the Curvature of Space

hypothetically in a continuous line, leaves no trace in the water, which, immediately the ship has passed, closes up behind, resuming its former static condition. If we come to the flight of an aeroplane moving through voluminal space (elaborately described by Einstein as a three-dimensional space-time continuum, with time added as a fourth space-dimension), we find exactly the same occurrence as in the case of a train or ship. The aeroplane describes a continuous line, which is merely imaginary and will leave no trace in the space through which it travels.

At this point Einstein and Hawking enter the field, complicating the story by bringing in gravity. An aeroplane, although apparently following a straight path, will, when flying parallel to the earth's surface over a long distance, describe a curved path, which, induced by the earth's gravitational pull, follows the round shape of our globe (a so-called geodesic). Worthy of note here is that the gravitational pull will be on the aeroplane, not on the curved path the plane will take, which delineates a purely imaginary curve and leaves no trace in space, just as a football kicked into the air will leave no trace in the space over the football field.

However, Einstein and Hawking insist that the curved path the aeroplane describes is a real physically perceptible curve, as if the plane was carrying a spaghetti-making machine, leaving a long continuously curved trail of spaghetti behind. This rather drastic example of the curvature of linear space simulates the idea of the two scientists, presenting the space curve of the aeroplane as if it were a solid substance.

So far we have dealt with examples wherein the so-called space continuum (whether straight or curved) is merely imaginary, and does not form a truly continuous line. The next two examples deal with what appear to be genuinely continuous space entities. The first is a light-

Has Hawking Erred?

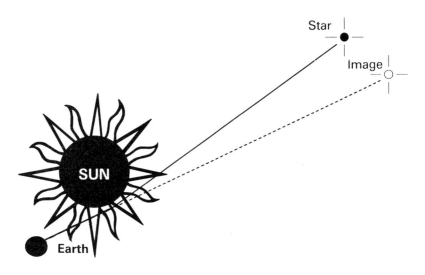

Fig. 7: Lightbeam passing the Sun

beam passing the sun (see fig. 7 above), the second a jet of water directed towards a burning house.

According to accepted theory, light moves both in the form of waves and in beams of particles which have mass. Einstein accordingly conjectured that light must be subject to gravitational pull. Thus the light of a distant star when passing a massive body like the sun could, induced by the sun's gravitational pull, be deflected from its straight path and form a bend. To test the phenomenon, one had to avoid the glare of the sun and arrange the test during a solar eclipse. When Einstein's assumption was tested under such conditions in 1919 his prediction proved correct. (See fig. 7 above).

However, no such lightbeam coming from outer space can be a permanent feature, as it is bound to shift or disappear from the earth's view as the earth moves along in space. As Hawking rightly remarks (H32):

As the earth orbits around the sun, different stars

The Enigma of the Curvature of Space

appear to move behind the sun and have their light deflected. They therefore change their apparent position relative to other stars.

The bending of light rays, which theory proves to possess mass and therefore to be capable of being deflected by gravity, can be simulated (at least in principle) by the example of a water jet directed towards a building on fire. Like the mass of the lightbeam the mass of the water jet describes a path curved by the earth's gravity. Both examples, lightbeam and water jet, present us with a continuous line, passing through what for all practical purposes can be described as empty voluminal space. When both cease to flow, neither lightbeam nor water jet will have left the slightest trace in the space through which they have passed.

Of further interest here is that a lightbeam, besides being subject to bending, also forms a continuous line. Expressed in Einsteinian terms it is a linear, one-dimensional space continuum, which, with the time factor added as a bonus, makes it a space-time continuum; although in all our previously submitted models of space continuums, such as a ruler, a rail track, a ship, an aeroplane, a lightbeam and a water jet, time did not intrude on, or intermingle with, the space factor, which belongs to a different category of phenomena. In the timing of events, time merely plays the part of an independent, comparative measure.

Referring further to the curvature of space and its credibility, Hawking (p. 31) turns the light example inside out, by writing that 'the fact that space is curved means that light no longer appears to travel in straight lines in space'; thereby making the bending of the light seemingly responsible for the curvature of space. The use of such arguments in explaining the curvature of space is further emphasised when Hawking (H33) describes time and

Has Hawking Erred?

space as dynamic entities. When forces act and bodies move they affect the curvature of time and space. In contrast, the structure of space-time influences the movement of bodies and the action of forces. Time and space not only influence but are themselves influenced by all the events that occur in the universe.

In further pursuit of this principle quite fantastic space-configurations were born. This finds expression in Hawking's book (H135). Commenting on the 'Origin and Fate of the Universe', Hawking refers to Einstein's idea that curved space-time is represented by the gravitational field, whereby particles try to follow the nearest thing to a straight path in space that is curved. However, as space-time cannot be characterised as flat, they describe a bent path as if being influenced by a gravitational field. With all respect for Hawking's other remarkable achievements, such arguments seem forced and convoluted. Yet Hawking uses this enigmatic phraseology to elevate space-time and its curvature to the status of a universal phenomenon.

The most remarkable omission in Einstein's and Hawking's reasoning is their failure to explain how the principles of space-time and of the curvature of space become applicable to three-four- (and more) dimensional voluminal space. All the examples they advance in this respect are squarely based on linear space projections (a ship, an aeroplane, a lightbeam or a water jet progress strictly linearly). Yet both scientists, suddenly and without further explanation, transfer the principle of linear space into voluminal space, although there appears nowhere any justification for this surprising transposition.

That Hawking has adopted the basic principle of **linear space** to explain space-time and subsequently applies it to both voluminal space and the curvature of space-time, finds confirmation in this book:

The Enigma of the Curvature of Space

Page 22: Under the space-time diagram, fig. 2.1, he writes,
'Time is measured vertically . . . the observer's path through space and time is shown as a **"vertical line"** ' [emphasis supplied].

Page 24: 'In relativity there is no real distinction between space and time co-ordinates'.

Page 25: In fig. 2.2, another space-time diagram, all the space co-ordinates are again given in terms of **linear space**.

Page 30: 'In general relativity, **bodies always follow straight lines** in four-dimensional space-time but nevertheless, appear to move along curved paths in our three-dimensional space' [emphasis supplied].

Page 31: 'The general theory of relativity predicts that the light cones of points near the sun would be slightly bent inwards by gravitational fields'. [Comment: **light cones thus projected proceed strictly linearly.**]

Page 33: 'In the general theory of relativity space and time are dynamic quantities: when a body moves, or a force acts, it affects the curvature of space and time . . . (to which I must remark that this space and time principle is **only applicable in relation to linear space**)' [emphasis supplied].

The same objections hold true in Einstein's case, where he explains space-time by means of linear space examples. There is a ship and a train example on B42, two further railway examples on B44 and B45, and others as well, **all of them strictly linear.**

The most important of these is Einstein's aeroplane

Has Hawking Erred?

example (B58), which attempts to simulate the space-time concept. He writes:

> The flight of an airplane from New York to Los Angeles can best be pictured in a four-dimensional space-time continuum. The fact that the plane is at latitude x, longitude y and altitude z means nothing to the traffic manager of the airline unless the time co-ordinate is also given. **So time is the fourth dimension** [emphasis supplied].

Let us analyse this example once more, firstly by referring to Einstein (also B58), who writes:

> An aviator guides his plane through a three-dimensional space continuum, hence he has to consider not only longitude and latitude, but also his height above the ground. The continuum of an airplane pilot constitutes space as we perceive it. In other words, the space of our world is a three-dimensional continuum.

What is obvious here is that when Einstein speaks of our earthly space being a 'three-dimensional space continuum', he defines a concept which equals our own definition of 'universal voluminal space'.

Also obvious from Einstein's example is that the three space co-ordinates, longitude, latitude and height, are all one-dimensional linear space projections, by means of which the position of the aeroplane is determined. It can be assumed that all the measurements are made in conventional metric units of length. These measurements, all of them linear one-dimensional, are completely indifferent to the universal voluminal space within whose bounds their measurements are carried out; the universal voluminal space remains unaffected by the act of measuring and plays an entirely passive part in the transaction. As to

The Enigma of the Curvature of Space

time, which is supposed to form an extra, fourth space-dimension, the following principle applies.

As already pointed out, space and time are two distinct phenomena, with space denoting distance and time denoting duration. Time cannot therefore play the part of an extra space-dimension, since the regular time flow recorded by timepieces (and related to the earth/sun cycle) is only casually related to the flight of the aeroplane. For example the same clock or watch which monitors the position of this particular aeroplane, records at the same time the minutes it takes to bring a pot of tea to the boil, as well as recording the pot's own ageing. In the two latter cases no space factor is involved. In the case of the tea, to use the Einstein rationale, one could well conjecture that time plays the part of an extra 'temperature dimension'; while in the case of the timepiece recording its ageing, the time factor remains entirely wrapped up in its own time-dimension.

In the light of such examples, which could be extended a hundredfold, Hawking's assertion (H23) that time is not separate from space, but combines with it to form an object called space-time, loses all credibility. Nor can we accept the Einstein assertion (also quoted by Hawking (H23) that 'In relativity there is no real distinction between space and time co-ordinates, just as there is no real distinction between space co-ordinates'.

We have seen that Hawking's and Einstein's space-time examples, which, besides pointing to the obsolescence of space-time, also relate to the curvature of space and time, are all squarely based on linear space projections, although both authors identify them subsequently with the properties of an assumed four-dimensional voluminal space continuum. Yet in accordance with what was demonstrated above, the principle of linear space in relation to time measurement cannot be imposed on three-dimensional voluminal space. As a non-material void this last

Has Hawking Erred?

can neither bend nor warp, nor can it be in any way affected by any object passing through it, be it an aeroplane, a light cone, a jet of water or a football. None of these or any other object can leave the slightest trace in empty voluminal space, either during or after its passage. The postulation of a fourth voluminal space-time dimension and its curvature, postulated apart from linear space, is therefore quite incompatible with the criteria of voluminal space as defined in this work.

16
Velocity of Light and the Solar Calendar

An interesting fact is the association of the speed of light with our solar calendar. This was already briefly referred to in chapter 12 in relation to 'Train example 3' and is linked to Einstein's particular interpretation of the Michelson-Morley experiment.

Barnett comments (B37):

> The Michelson-Morley experiment confronted scientists with an embarrassing alternative. On the one hand, they could scrap the ether theory which had explained so many things about electricity, magnetism and light. Or, if they insisted on retaining the ether, they had to abandon the still more venerable Copernican theory that the earth is in motion.

Einstein's assessment of it reads (B38): 'The one indisputable fact established by the Michelson-Morley experiment was that **the velocity of light is unaffected by the motion of the earth**' [emphasis supplied].

According to Barnett, Einstein conjectured further that:

> the velocity of light is constant regardless of the motion of the sun, moon, star, meteor or other systems moving anywhere in the universe. From this, Einstein drew his

broader generalization, asserting that the laws of nature are the same for all uniform moving systems.

Einstein's failing here is based on his unconditional acceptance of the dictum **the velocity of light is unaffected by the motion of the earth**. The ambiguity in this statement rests on the consideration that the velocity of light, or rather its measurement, while not directly affected by the motion of the earth, is nevertheless theoretically linked to the motion of the earth.

This thesis rests on the contention endorsed by Einstein himself that the speed of light is measured at 186,284 miles per 'second'. The second referred to is our own earthly second, being part of our earthly time standard. Its duration relates to the fraction of the time the earth needs to move once around its own axis in one day. The entire day again totals 86,400 terrestrial seconds. In other words, the measuring of the velocity of light rests on our earthly time system, identical with what we have defined as 'regular calendar time'. The earthly linkage to the speed of light is further emphasised by the fact that the distance the light travels in one second i.e. 186,284 miles (or about 300,000 km) is based on our earthly metric system (expressed above in the equivalent of miles), which is again based on our earthly standard metre. The latter's length is synchronised to fractions of an earthly second, accurately measured by cesium clocks.

These examples show that the speed of light is, in the first place determined or measured relative to the motion of the earth. However, if we abandon the relativity principle and use the velocity of light as an independent absolute constant, unrelated to the earth or any other heavenly body, then calendar time must also become a universal constant. In other words, the postulate of the absolute speed of light is intrinsically associated with that of 'regular calendar time', elevating the latter to another

Velocity of Light and the Solar Calendar

universal absolute in terms of which the duration of all events in the universe can be measured. In other words, conventional calendar time can be **conditionally** put on par with 'absolute time' as previously defined by Einstein – though rejected by both Einstein and Hawking. It was Einstein's persistent rejection of our terrestrial time-standard, itself identical with 'regular calendar time', which became a constant irritant in his own endeavour to reconcile what he describes as the seemingly irreconcilable conflict between his belief both in the constancy of the velocity of light, and in the principle of the addition of velocities; although, as he points out, 'the latter appears to **rest on the stern logic of mathematics i.e. that two plus two always make four**' [emphasis supplied]. Einstein has described this simple equation as a fundamental law of nature (B43).

This apparent conflict between the two concepts prevalent in mathematics raises doubts that the use of mathematics in physics is not always reconcilable with nature's manifestations. Physics' unconditional reliability on mathematics can by no means be taken for granted (see chapter 3).

17
Other Observations

My object in questioning certain aspects of Hawking's book (as well as drawing attention to flaws in Einstein's reasoning), has been twofold. Firstly, to emphasise the role which regular calendar time plays in physics, insofar as all events happening in the universe are chronologically aligned to it and measured by it (there has never been any alternative time measurement in modern physics). Secondly, to show that the concept of space-time associated with the curvature of space and time is a fallacy which has no basis in physical reality.

Among other theories, that of the curvature of space has mainly been derived from considerations concerned with the discovery that light rays can curve, or be deflected, by gravitation. We have shown that this principle is based on a strictly two-dimensional linear space curvature which cannot be imposed on three-dimensional voluminal space. Nevertheless, basing themselves on the latter supposition, some physicists have speculated that the possible shape of the universe could be imagined as a geodesic, being finite in extent and without boundaries. For example, Barnett (B84) argues that 'in the Einstein universe there are no straight lines, there are only great circles. Space in it, though finite, is unbounded'. It must be imagined as having the form of a sphere. Hawking

Other Observations

(p. 42) explains that in an expanding universe the galaxies are moving directly away from each other and compares the situation to an expanding balloon on whose surface spots have been painted. As the balloon is inflated, the distance between the various spots grows larger, while at the same time their movement apart will become faster. However, none of the spots can be designated as the hub of expansion.

The astronomer Edwin Hubble (B85–86), working with Einstein's field equations, calculated that the radius of the universe measures 35 billion light years. At the same time he stated that the years in question are expressed in terms of terrestrial calendar time. Moreover, Barnett (B86) claims that:

> Einstein's universe, while not infinite, is nevertheless sufficiently enormous to encompass billions of galaxies, each containing hundreds of millions of flaming stars . . . A sunbeam setting out through space at the rate of 168,284 miles per second, would, in the universe, describe a great cosmic circle and return to its source after little more than 200 billion terrestrial years.

This hypothesis contains however, one major drawback: While Einstein proved (see fig. 7) that light rays can be deflected by gravity from their straight path, when passing a massive star like the sun, he omitted to mention that the light rays, having passed this first massive body, may, when continuing on their path, encounter further massive bodies, or suns, which would induce them to bend again and again. This kind of suggested chain reaction is illustrated in figs. 8, 9 and 10.

Has Hawking Erred?

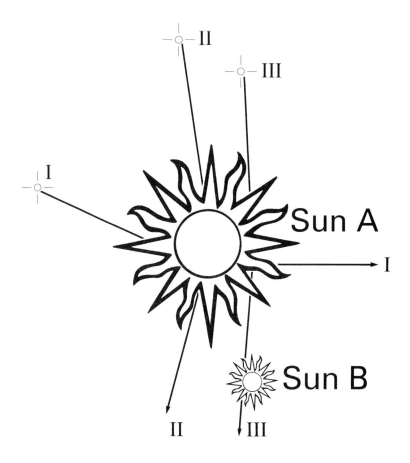

Fig. 8: Paths of Light from Stars

Figure 8 shows the path of light from different stars, I and II, as their rays pass sun A. The light of star III is again deflected by another sun, B, which may be millions of light years distant.

Other Observations

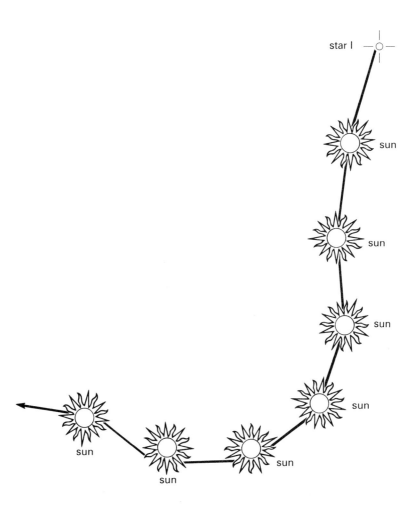

Fig. 9: Passage of Light from Stars

Figure 9 shows light of star I passing several suns, each separated from the other by millions of light years.

Has Hawking Erred?

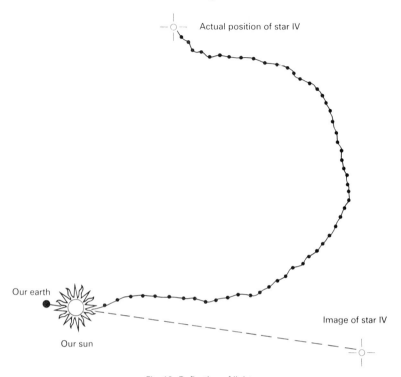

Fig. 10: Deflection of light

Figure 10 shows light from a very distant star iv, possibly billions of light years distant from our sun, experiencing many deflections

In commenting further on fig. 10 I would point out, as Einstein does, that the universe encompasses billions of galaxies, each again containing billions of suns. The light rays from any star lodged near the universe's fringes will, while passing through the universe, encounter countless suns (indicated by little black spots in fig. 10), each of which, as the light rays pass it, will deflect and curve the path of these rays. Hence, when an observer on earth finally receives the light of star IV, he will place it in the imaginary position indicated in fig.10, which vastly differs from its real position in space.

Other Observations

I suggest therefore that Hubble's assumption that a light ray setting out from a star will return to its original position after having described a full circle around a geodesically shaped universe, may, in view of the above considerations, no longer be sustainable. Neither may the hypothesis of a geodesically shaped, self-contained and finite universe be any longer sustainable, if light rays can move irregularly in the manner indicated in fig. 10, a fact which also conflicts with the curvature of space syndrome proposed by Einstein and Hawking.

For example, Hawking writes (H33):

Space and time are now dynamic quantities; when a body moves, or a force acts, it affects the curvature of space and time – and in turn the structure of space-time affects the way in which bodies move and forces act. Space and time not only affect but also are affected by everything that happens in the universe.

And further (B74):

Just as a fish swimming in the sea agitates the water around it, so a star, a comet or a galaxy, distorts the geometry of the space-time through which it moves.

This means that if any object, including light rays, could influence the curvature of space, as both Hawking and Einstein claim, then all the innumerable erratically travelling light beams criss-crossing the universe since its beginning would have scrambled up universal voluminal space into such an entangled Gordian knot that even a heavenly Hercules could not have unscrambled it. It stands to reason that in such a spatial entanglement the movements of the heavenly bodies would long ago have come to a standstill. However, physical reality proves that neither light rays nor any other force, including gravity, can

Has Hawking Erred?

influence universal voluminal space, which being merely an abstract hypothetical void and lacking any material existence, cannot be influenced by physical occurrences of any kind. The proof is the existence of the universe as we find it today, unensnarled by the acrobatics of space-time and space-curvature.

18
Commenting on the Big Bang

A recent discovery seems to confirm the view long held by some of the world's leading physicists that the 'Big Bang theory' so far offers the best explanation of how the universe may have come into being. The details were announced at a meeting of the American Physical Society, held in Washington DC on 23 April 1992. They seemed to confirm previous assumptions that about 15 billion years ago a tiny ball of concentrated mass (with no identifiable source) exploded in a 'big bang' and in expanding created the universe as we find it today. The discovery was made by the Cosmic Background Explorer – COBE – spacecraft of the USA National Aeronautics and Space Administration, whose Space Flight Center's computer recorded blotches of all sizes, indicating a region in the universe where the background radiation microwaves are a minuscule 30 millionths of a degree warmer than average. This almost imperceptible variation in temperature, it is asserted, is the missing link now confirming the Big Bang theory.

Also registered were wispy wrinkles in giant clouds of microwave background, which are estimated to have appeared about 300,000 years after the Big Bang. Earlier instruments, less accurate, had shown that the microwave background was incredibly smooth, suggesting that particles could not have clumped together to form clusters of great galaxies. Hawking, who had long predicted the possibilities now revealed by COBE, hailed this as the scientific event of the century. Carlos Frenk, a physicist at Durham University (Reuter, London, 25.4.92), commented further:

> The discovery confirms scientists were correct in looking for another missing link in explaining the universe's dark matter. The matter that scientists have been able to find – stars, galaxies and other formations – would not explain why there is so much observable gravity. Gravity, the force that keeps people from flying off the surface of the earth and which keeps the earth in orbit around the sun, is produced by matter.

The scenario suggested by the findings of the COBE spacecraft summarised in the following five points:
1. The Big Bang theory states that all matter was concentrated in a tiny pinpoint, smaller than the finest speck of dust, infinitely dense and infinitely hot, until it was released in a mighty cosmic explosion 15 billion years ago.
2. During what is called the inflationary period the universe between an age of 10^{-35} of a second and 10^{-33} of a second, expanded from the size of less than an atom to the size of a grapefruit.
3. Within a period of between three minutes after the Big Bang to 300,000 years after the Big Bang, the universe expanded into a fog of free-moving charged particles. During the end of it the expanding cosmos cooled and the particles combined into atoms. Some areas were denser than others, forming gargantuan gas clouds.
4. In the period of between 300,000 years and 2 billion

Commenting on the Big Bang

years after the Big Bang, under the influence of gravity, the colossal clouds gradually broke up into smaller, galaxy-sized structures.
5. Between 2 billion years ago and the present (15 billion years after the Big Bang), stars formed into galaxies, while the universe continued to expand.

To illuminate the forces which constitute the preconditions immediately before the Big Bang we have to delve into particle physics.

According to Hawking (H66), within the existing wave/particle duality not only can light and gravity be described in terms of particles, but equally so can everything else that exists in the universe. Furthermore, particles possess a property called spin. According to the quark theory all matter consists of quarks or leptons, which are the only true elementary particles; however, no quarks have yet been identified experimentally. Hawking points out that twenty years ago it was thought that quarks were the ultimate elementary particles, meaning that they were indivisible. Now it has been found that quarks are possibly made from still smaller particles.

There now exists evidence which strengthens our belief that we are near the discovery of the ultimate substance from which the universe is made. For example, neutrinos (H185) are extremely light (possibly massless) elementary matter particles which are affected only by the weak force of gravity. On our earth, matter is mainly composed of neutrons and protons, which again consist of quarks. There is also evidence from cosmic rays that the same is true for all the matter in our galaxy. Although there is no direct evidence, it can be assumed that matter in other galaxies is also made up of quarks.

The presumed shaping of the universe
How does Hawking and other supporters of the Big Bang theory imagine the universe was shaped during its

Has Hawking Erred?

assumed 15 billion years of existence? The view is that before the Big Bang there existed neither time, nor space nor matter, meaning that before the birth of our universe there existed absolutely nothing. In this presumed zero scenario a tiny speck of mass appeared 15 billion years ago, and in its Big Bang explosion expanded into our universe. No indication is given how this speck of matter was born.

In order to accommodate the curvature of space and time in the creation of the universe, Hawking compares (H137) the possible shape of the universe with the round-shaped surface of the earth. As mentioned before, Hawking (H42) proposed that the inflation of the universe can be compared with the blowing-up of a balloon with a number of spots painted on it (see p. 135).

With the balloon expanding the distances between any two spots increased. But none of the spots could be considered to be the centre of expansion. The situation is summarised by Hawking (H45) by referring to three models proposed by Friedmann. In the first model of the universe, which expands and recollapses, space is bent in on itself, like the surface of the earth, and therefore is finite in extent. In the second model, the ever-expanding one, space is bent the other way, like the surface of a saddle, making its space infinite. In the third Friedmann model space is flat and therefore also infinite.

One example of this is perhaps Hubble's model of the universe (B85–86), which he estimated as having a radius measuring 35 billion light years. This distance measured from the centre would mean that the universe was 35 billion years old, while a lightbeam encircling it would need over 200 billion years to return to its starting point.

However, neither Hawking's nor Hubble's model would resemble the solid body of the earth, but would merely consist of the surface enveloping a hollow interior. The model would resemble the hollow peel of a giant

Commenting on the Big Bang

orange. This somehow coincides, at least graphically, with the spherical sheet postulated by Einstein and Hawking, representing curved space-time. The dilemma of how this space-time sheet of unstated thickness can accommodate all the heavenly bodies and their movements is solved by the Delphic assertion among Hawking's final conclusions (H173) 'that space and time may form a finite, four-dimensional space without singularities or boundaries, like the surface of the earth, but with more dimensions'. By postulating the possibility of more than three space-dimensions for voluminal space, a proposition which defies all bounds of human understanding and can therefore not be rationally contemplated, Hawking gets away with his curved space-time myth.

The above attempt to explain how the universe formed and expanded leaves several questions unresolved. For example, the Big Bang is supposed to have created not only the matter of the universe but also all its space. It is maintained that before the event neither matter nor space existed anywhere. As presented by Einstein and Hawking, the space of the universe is identifiable with a gravitational field, which contains all heavenly bodies, including dark matter, and forms a curved surface shaped like the earth. The process of creation is also compared with the blowing up of a slowly expanding balloon (its surface compared to a curved space-time sheet), on the surface of which constellations of stars and other matter slide slowly apart like ice-skaters.

This raises two important questions. If this curved space-time sheet is shaped like a sphere, forming an envelope of unknown thickness, it must be hollow inside, presenting a huge volume of empty space, whose presence is totally ignored by Hawking. On the other hand, this assumedly empty space (not mentioned in the theory), cannot be part of the gravitational field postulated by Einstein and Hawking, who present the latter as a

Has Hawking Erred?

curved space-time sheet. Whether the theory can cope with this new and intricate problem hovering behind a curtain of space-time jargon, is highly questionable.

The second point to be considered is that the curved space-time sheet of unspecified thickness must be supposed to contain all the existing heavenly bodies in the universe, including of course all the galaxies. Hawking describes our own slowly rotating galaxy as measuring about one hundred thousand light years across, our galaxy being merely one of some hundred thousand millions of galaxies that have been discovered by astronomers – some of them containing some hundred thousand million stars. To imagine that this colossal amount of rotating galaxies, filling a space measuring millions of light years across, can be squeezed into a round-shaped space-time sheet resembling a global envelope – whose internal space volume is condemned to total oblivion by the theory – is a proposition defying normal intelligence. Neither does it match, in structural respects, the real-life observations of contemporary astronomy.

A further scrutiny of the Big Bang

Hawking (B117) states: 'At the Big Bang itself the universe is thought to have had zero size and so to have been infinitely hot'. Strictly interpreted, the phrase 'zero size' means non-existence, while on the other hand, something that is infinitely hot must have a material existence. Consequently, Hawking's expression 'zero size' needs to be amended to read 'almost zero size'. Hawking also states that one hundred seconds after the Big Bang the temperature would have fallen to one thousand million degrees. According to him (H116), 'temperature is simply the measure of the average energy or speed of particles'.

To measure the energy embodied within the particles forming the tiny speck of matter from which the Big Bang exploded, it must be presumed that they spun at an incredibly furious pace to reach the super-temperatures claimed. The amount of heat energy available can be obtained by employing Einstein's formula $E = mc^2$. Accordingly we have to multiply the combined mass of the available particles at the time of the Big Bang with the square measure of the speed of light (which is 300,000 km per second), obtaining a figure of 90,000,000,000 (i.e. 90 billion) km per second. When multiplied with the mass of the particles available at the Big Bang (which, according to theory, ought to be equivalent to the total mass of particles in the present universe) we get the total energy available. Barnett (B55) writes: 'Einstein showed that mass and energy are equivalent and that the property called mass is simply concentrated energy. In other words, matter is energy and energy is matter and the distinction is simply one of temporary state'.

What actually happens in nature is that in every case of energy conversion a specific mass of energy-producing particles is converted into another type of mass, again producing energy. In the process of such conversion there are produced the photons of light, the gravitons of gravity, the quarks, and any or all other types of particles that populate the universe and constitute the source from which atoms and molecules, and eventually all heavenly bodies which make up the present universe (including animals and humans), have derived their existence.

All this conforms with the law of the conservation of energy, which according to Hawking (H184) is 'The law of science that states that energy (or its equivalent in mass) can be neither created or destroyed'.

Other aspects of the COBE discovery

The COBE discovery which consolidates claims that the universe was born in a Big Bang also reveals two important aspects of time. One is that the five-point scenario previously detailed (p. 142) inadvertently endorses the

Has Hawking Erred?

concept of absolute time as defined by Einstein. The second is that the exclusive use of conventional calendar time in the above mentioned scenario clashes with the assumptions of the space-time curvature syndrome advanced by Einstein and Hawking, meaning that we have here two differing time concepts. As shown below these cannot be reconciled.

The first aspect of time:
The COBE discovery reveals among other things that between two immensely short time intervals of less than a split second, namely between 10^{-35} and 10^{-33} seconds after the Big Bang, the universe expanded from a size of less than an atom to the size of a grapefruit. In the subsequent expansion of the universe differing events are measured in the stated intervals of two minutes, 300,000 years, two billion years, five billion years and fifteen billion years. I repeat my previous conclusions that the time measurement employed in this process is strictly based on the regular flow of conventional calendar time, which is ultimately linked to the earth/sun relationship. In our above scenario this very time concept is retroactively employed back to the very birth of the universe to measure events lasting the tiniest part of a split second. And all this despite the fact that regular calendar time was only invented about fifteen billion years after the assumed birth of the universe. In short, we have here an example of the practical use of what Einstein (B40) had defined as 'absolute time' (a definition later endorsed by Hawking), namely '**A steady unvarying inexorable time flow, streaming from the infinite past to the infinite future**' [emphasis supplied].

The 'infinite past' reaches of course back to the birth of the universe, as postulated by Einstein and Hawking. It stands to reason that this definition does not allow variations in the flow of time to occur (much less allow a

Commenting on the Big Bang

standstill of time) anywhere in the universe during its time-measured existence, nor is the time flow influenced by spatially related considerations. The fact that Einstein indicated that the acceptance of 'absolute time' (as defined by him and now endorsed in the chronology of the COBE spacecraft findings) would contradict the assumptions of his special relativity renders the latter theory open to doubt.

The second aspect of time:
We will first quote from Hawking on the space-time curvature and then contrast it with the flow of regular calendar time. According to Hawking (H133), 'The beginning of time would have been a point of infinite density and **infinite curvature of space-time**' [emphasis supplied]. One feature which Hawking (H135) believes must be part of any ultimate theory is the idea advanced by Einstein that the gravitational field is represented by **curved space-time**.

According to theory, the Big Bang produced both the mass and the space of the universe, the latter in Einstein's opinion representing a gravitational field, which can be described as a **'complete curved space-time'**! As to the function of time within **curved space-time**, the question is whether it can be reconciled with the regular flow of conventional calendar time, whose sole use is that of a comparative measuring device.

Before trying to answer this question it is necessary to reconsider and add to the definition of 'time' already elucidated in chapters 5; 6 and 7 of this work. For example, the abstract word 'time' used in a phrase 'the beginning of time' is merely a popular generalisation of something intangible which has no scientific standing. 'Time' or the time concept can only find practical application in everyday life or science if used as a comparative measure. This is because all occurrences in the universe take place irres-

Has Hawking Erred?

pective of whether their duration is measured (i.e. timed) or not. Nor does the application of time measurement change the course of events or affect the events themselves. It is therefore suggested, although this may raise the eyebrows of many physicists, that time has no influence on the course of events. What this further means is that **all events in the history of the universe, would have happened in exactly the same order even if there had been a total absence of what we define as 'time'**. In other words, 'time' has no bearing on events. It is a purely human invention serving solely as a comparative measure attuned to human needs both practical and scientific. It has no other function.

In contrast the Hawking-Einstein time concept, as in the 'curvature of space-time', not only implies a fusion of time and space (linear as well as voluminal) but means a mingling of two entirely different phenomena: one, 'time', an abstract and non-material entity, the other, 'space', having a distinct though strictly hypothetical presence. It also means that 'time', now becoming part of the new mix of space-time, also assumes a material presence within this new combination which can bend and curve. This new property of time, namely curvature, is confirmed by Hawking above (H135) when he speaks of a gravitational field being represented by curved space-time. **There is no indication that within this particular system of curved space-time, time can retain its measuring function**.

In contrast, if we consider the flow of regular calendar time over the last 15 billion years (the assumed lifetime of the universe), as evidenced in the five-point COBE scenario, we can see no possibility that the regular time flow demonstrated by it can be subjected to any bending or curving; nor can it be otherwise manipulated to conform to the Einstein-Hawking theory of curved space-time. The conclusion is therefore that ordinary conventional calendar time cannot be reconciled with the time principle embodied in the Einstein-Hawking idea of a curved space-time.

The fact that Einstein and Hawking make exclusive use of conventional calendar time when timing astronomic events makes their idea of a curved space-time (lacking any time measuring function) not only appear contradictory, but actually aborts its right to existence.

Other aspects of the Big Bang

The Big Bang theory proclaims that the universe began when a dense knot of matter-energy occupying a space of near zero size erupted in a mighty explosion about 15 billion years ago.

According to scientific dictionaries, the 'law of the conservation of mass and energy' proclaims that the sum total of energy and mass multiplied by c^2 (where c is the velocity of light) is constant for any system and cannot increase or decrease, nor can it be created or destroyed. While not disputing the indestructibility of mass and energy and their unchangeable quantities prevalent in the universe, we need to recall that mathematically expressed in Einstein's formula $E = mc^2$, 'energy' and 'mass' are not the same as previously mentioned and 'E' (standing for energy) does not equal 'm' (standing for mass), as this would read $E = m$. Instead, according to Einstein's formula, $m = E/c^2$, hence 'E' = energy is divided by c^2. On the other hand, a total conversion of mass into energy is theoretically possible by multiplying mass, expressed in grams, kilograms or tons, by c^2, thereby transforming mass into kinetic energy.

In compliance with the above law of mass and energy conversion, the total mass or matter (expressed in units of weight) prevalent in the universe must have originally equalled the amount of matter (also expressed in units of weight) which was encapsuled in the near-zero-sized pinpoint of matter, the existence of which must be pre-

Has Hawking Erred?

sumed to have preceded the Big Bang. If this were not so it would have failed to produce the same amount of mass or matter which the universe presently contains.

We will now examine the premise whether the entire physically available mass or matter in the universe could have been produced from the tiny primordial wisp of mass which exploded in the Big Bang.

To make a rough estimate of the total of physical mass prevalent in the universe we will reduce it to the mass, equivalent in weight, residing in the number of suns estimated to equal the universe's total mass or weight. The weight of our own sun will serve as a prototype. As to the estimated number of suns comprising the total mass of weight in the universe, the Hubble telescope was reported (*International Herald Tribune*, 4 June 1992) to have detected a new class of objects in the universe which were gigantic star-forming clusters created from the wreckage of two colliding galaxies situated at a distance of 230 million light years from our planetary system.

> The star-forming clusters put out energy so intense that collectively it is equivalent to 500 billion suns. That output exceeds that of a supermassive black hole equal to about 400 billion suns believed to be hiding nearby at the core of the Arp 220 galactic debris.

I also refer to a previously mentioned observation by the astronomer Hubble himself, who wrote (B86) that 'Einstein's universe, which is not infinite, is sufficiently enormous to encompass billions of galaxies'.

Based on this estimate, which is very vague as to the exact number of galaxies, I will, for the purpose of practical calculation, assume the figure of five billion galaxies, though there could be more. As to the average number of suns in each galaxy I base my figure on the above article and assume an average of 400 billion suns per galaxy.

Commenting on the Big Bang

Included in their mass are all the billions of planets, black matter and black holes, clouds of particles and radiation and any other physically perceptible matter which the galaxy may house. This gives us a total number of suns representing the content of matter in the universe at a conservatively estimated mass equivalent of 20,000 billion suns, (taking the weight of our own sun as a convenient prototype and unit). It has been estimated that the cubic content of our sun amounts to 16,000 thousand trillion cubic metres. By arbitrarily assuming that the weight of one cubic metre of solar mass (having one quarter of the earth's density) equals one quarter of a metric ton of weight we obtain a total of 4,000 trillion tons. Multiplying this number by 20,000 billion (the amount of sun-mass equivalent in the universe) we get **the total weight of matter in the universe, which amounts to 80,000,000,000,000 trillion tons or 8, followed by 25 zeros.**

The idea that a speck of matter smaller than a dust particle on my table could have accommodated the condensed mass or weight of the entire universe, equalling 20 billion suns, stretches credibility beyond limits. It is a fantasy lacking genuine scientific backing. It is for this reason, and others previously mentioned, specifically the involvement of the space-time factor, that the Big Bang theory as it now stands must be seriously questioned. A review of its basic assumptions is certainly indicated.

19
Concluding Remarks

I would like to conclude the book by reiterating what has been its guiding principle throughout. It is that **it is absolutely axiomatic and incontestable that all events in the universe from its presumed beginnings between 10 to 20 billion years ago up to the present would have happened in exactly the same way, whether they are subjected to present or retroactive time measurement or not**. In short, 'time', or the process of 'time measurement', exercises no influence whatsoever on physical events, just as the metric measurement of distances does not affect the objects measured, or the measurement of temperature the objects of such measurement.

Just as the process of measuring temperature or distance does not affect the voluminal space within which such measuring takes place, so does time-measuring not influence the voluminal space within which the duration of time between events is measured or assessed.

The acceptance of this principle further precludes any possibility of a fusion or intermingling between the concepts of time and space, as the two belong to entirely different classes of natural phenomena. Such a fusion being thus unacceptable I feel justified in describing the space-time concept as a major scientific fallacy.